Oil Companies International Marine Forum

The Safe Transfer of Liquefied Gas in an Offshore Environment

(STOLGOE)

First Edition - 2010

The OCIMF mission is to be the foremost authority on the safe and environmentally responsible operation of oil tankers and terminals, promoting continuous improvement in standards of design and operation

Issued by the

Oil Companies International Marine Forum

29 Queen Anne's Gate
London
SW1H 9BU
United Kingdom

First Published 2010

ISBN: 978 1 85609 400 9
eBook ISBN: 978 1 85609 410 8

British Library Cataloguing in Publication Data
A catalogue record for this book is available from the British Library.

The Oil Companies International Marine Forum (OCIMF)

is a voluntary association of oil companies having an interest in the shipment and terminalling of crude oil and oil products. OCIMF is organised to represent its membership before, and to consult with, the International Maritime Organization and other governmental bodies on matters relating to the shipment and terminalling of crude oil and oil products, including marine pollution and safety.

Terms of Use

The advice and information given in this 'The Safe Transfer of Liquefied Gas in an Offshore Environment' is intended purely as guidance to be used at the user's own risk. Acceptance or otherwise of recommendations and/or guidance in this Guide is entirely voluntary. No warranties or representations are given nor is any duty of care or responsibility accepted by the Oil Companies International Marine Forum (OCIMF), the membership or employees of OCIMF or by any person, firm, corporation or organisation (who or which has been in any way concerned with the furnishing of information or data, the compilation or any translation, publishing, supply or sale of the Guide) for the accuracy of any information or advice given in the Guide or any omission from the Guide or for any consequence whatsoever resulting directly or indirectly from compliance with, adoption of or reliance on guidance contained in the Guide even if caused by a failure to exercise reasonable care on the part of any of the aforementioned parties.

Witherby Seamanship is a division of Witherby Publishing Group Ltd

Published in 2010 by
Witherby Publishing Group Ltd
4 Dunlop Square
Livingston,
Edinburgh, EH54 8SB
Scotland, UK

Tel No: +44(0)1506 463 227
Fax No: +44(0)1506 468 999
Email: info@emailws.com
www.witherbys.com

Printed and bound in Great Britain by Bell & Bain Ltd, Glasgow

Purpose and Scope

These Guidelines address the transfer of Liquefied Petroleum Gas (LPG) between Floating (Production) Storage and Offloading facilities (F(P)SOs) and conventional gas tankers. Offshore operations involving the transfer of Liquefied Natural Gas (LNG) are not specifically addressed in this document, but will be included in a future update once industry experience is available on which to base guidance.

The Guidelines are primarily intended to familiarise Masters, ship operators, F(P)SO operators and project development teams with the general principles and equipment involved in LPG offloading activities between F(P)SOs and gas tankers. For the purposes of these Guidelines, the gas tanker is considered to be a vessel equipped with standard mooring and cargo transfer equipment, with a conventional single propulsion system comprising of a fixed blade propeller.

The guide primarily addresses side-by-side (SBS) operations where the gas tanker moors alongside the F(P)SO. LPG is also offloaded to gas tankers offshore in other modes, such as via an SPM or in a tandem configuration. These operations are considered in depth in other OCIMF publications associated with oil transfers. Within this publication, only the differences associated with gas transfer have been highlighted.

To address site-specific aspects in particular locations, the recommendations contained within this publication may be supplemented by additional requirements from development project teams, individual ship owners (or ship managers) and individual F(P)SO operators.

Individual countries may have local regulations that address mooring arrangements and LPG cargo transfer operations within their territorial waters or Exclusive Economic Zone. These must be complied with and, where appropriate, these Guidelines may be referenced for additional advice. These Guidelines should not be considered contrary to, or a replacement of, any National or International Regulations or individual terminal operator's specific procedures.

This publication primarily addresses the inter-relation between the F(P)SO and conventional gas tankers operating in SBS mooring configuration. The guidance includes recommendations for mooring equipment, considers mooring loads and operations, motions of the F(P)SO and gas tanker, station keeping, cargo transfer equipment and cargo transfer operations. The guidelines are intended to assist in standardising equipment and procedures for SBS activities and to highlight general design issues in the interest of safe, efficient and reliable operations.

Contents

Appendices 87

Glossary of Terms and Abbreviations

Assistant Mooring Master The person responsible to the Mooring Master for coordination of the mooring operation on the gas tanker and for connection of the transfer hoses/arms to the gas tanker's manifold. Job title and job description may vary between F(P)SOs.

CCTV Closed Circuit Television

Competent Person A person who has acquired through training, qualification and experience the knowledge and skills necessary to undertake the tasks that they are required to perform within their job description.

DPLT Dynamic Positioning Loading Terminal. A novel dynamic positioning offloading system that works on a friction/suction securing system to the offtake gas tanker. The system is an asymmetrical 'L' shaped unit that is deployed from the F(P)SO and is driven over to an offtake tanker. A hose allows product to be discharged from the F(P)SO to the DPLT unit and the discharge from the DPLT unit is either via hoses (oil) or loading arms (gas). The DPLT holds the offtake tanker at a safe distance from the F(P)SO without the need for mooring hawser or support tugs.

ERC Emergency Release Coupling

ESD Emergency Shutdown System

FPSO Floating Production, Storage and Offloading facility. This is a floating facility (usually ship or barge shaped) that receives, processes and stores hydrocarbons and has an offloading system to transfer treated hydrocarbons to conventional tankers.

F(P)SO Floating (Production) Storage and Offloading facility. A term used to refer to both FPSOs and FSOs in this document.

Fractionator A process for separating the gas components contained in LNG wherein a small amount of heavier reflux liquid (propane, butane, pentane, iso-butane and iso-pentane) is recycled to the top of a demethanizer column.

FSO Floating Storage and Offloading facility (usually ship or barge shaped) that receives and stores treated hydrocarbons and has an offloading system to transfer treated hydrocarbons to conventional tankers.

Gas Tanker A gas tanker equipped for regular trading and not specially designed or adapted for loading at offshore terminals requiring specialised mooring or bow loading equipment. In these Guidelines, the term 'gas tanker' is used to describe a vessel receiving LPG from an F(P)SO.

Hazid Hazard Identification Study

Hazop Hazard and Operability Study

Hindcast data Wave data derived from wind field information using a variety of methods ranging from simple formulae estimating windspeed, fetch and duration, to numerical models for translating wind field time histories covering large areas into corresponding wave field information. Large numbers of numerical wave prediction models are available for various areas of the world.

Insulating Flange A flanged joint incorporating an insulating gasket, sleeves and washers to prevent electrical continuity between ship and shore.

LNG Liquefied Natural Gas is predominantly methane compressed and cooled to minus 162°C to facilitate transportation in its liquid state.

LPG Liquefied Petroleum Gas. Propane, butane or a mixture of both. Propane is liquid at temperatures of less than approximately minus 42°C at atmospheric pressure. Butane is liquid at temperatures of less than approximately minus 7°C at atmospheric pressure.

MBC Marine Breakaway Coupling. A device designed to provide protection to the cargo transfer system and floating hose against surge pressures and/or axial tension in the hose by automatic shut-off of liquid flow and separation before the floating hose integrity is damaged.

MBL Minimum Breaking Load

MBR Minimum Bend Radius of the export hose. The minimum radius to which the hose may be bent without damage, permanent ovalling or deformation. Normally expressed as a factor of the hose internal diameter.

Mooring Master The person responsible to provide pilotage and mooring advice to the Master of the gas tanker, and for coordinating tug assistance and loading activities. The representative of the F(P)SO aboard the gas tanker during transfer operations. Job title and job description may vary between F(P)SOs.

MSDS Material Safety Data Sheet. A document identifying a substance and all its constituents providing the recipient with all the necessary information to manage the substance safely.

OGP International Association of Oil and Gas Producers. The representative organisation for International Oil and Gas Producing Companies.

OIM Offshore Installation Manager

Primary Fenders Primary fenders are large fenders capable of absorbing the impact energy of berthing and wide enough to prevent contact between the F(P)SO and gas tanker should they roll while alongside one another.

QC/DC Quick Connect/Disconnect Coupling

QRH Quick Release Hook

Safety Case A structured argument, supported by a body of evidence that provides a compelling, comprehensible and valid case that a system is safe for a given application in a given operating environment. Formal Safety Cases may be a locally legislated requirement.

Secondary Fenders Secondary fenders are small fenders used to prevent contact between the F(P)SO and gas tanker. They are particularly effective when rigged towards the end of a ship and are of most benefit during mooring and unmooring operations.

Side-by-side (SBS) Side-by-side (SBS) cargo transfers differ from ship-to-ship (STS) operations in that one of the vessels is a permanently moored facility (not underway), e.g. an F(P)SO, whereas the other is a gas tanker capable of ocean voyages.

SIGTTO Society of International Gas Tanker and Terminal Operators

SIMOPS Simultaneous Operations

SOLAS International Convention on the Safety of Life at Sea

Soliton An isolated wave that propagates without dispersing its energy over large areas, which can cause a rapid change in heading of the gas tanker and F(P)SO as the wave contacts the vessels in the form of a strong current.

SPM Single Point Mooring system. An integrated mooring arrangement for bow mooring a tanker. An SPM includes CALM (Catenary Anchor Leg Moored) buoy (also known as a Single Buoy Mooring (SBM), SALM (Single Anchor Leg Mooring), F(P)SO and a turret type mooring to a spar or similar structure.

Spread Moored An F(P)SO moored at a fixed location with its heading permanently aligned into the predominant weather/environmental forces.

Swivel A system that allows a flow of medium (fluid, electrical, air) from a fixed part to a rotating part. A swivel can be either an inline or a toroidal in configuration.

SWL Safe Working Load. A load less than the yield or failure load by a safety factor defined by a code, standard or good engineering practice.

Tail A short length of synthetic rope attached to the end of a mooring line to provide increased elasticity and also ease of handling.

Tandem Mooring Mooring two vessels in longitudinal alignment bow-to-bow, stern-to-stern, or in any fore-and-aft configuration.

Terminal Operator The operator of the F(P)SO responsible for the safe operation of the F(P)SO. The operator is not necessarily the owner of the F(P)SO or owner of the field (concession holder) and could be a contractor operating the F(P)SO on behalf of an oil company.

Turret Moored An F(P)SO moored at a fixed location, but free to align (weather vane) with the prevailing weather conditions. The turret can be internal or external.

Bibliography

Reference 1	*International Safety Guide for Oil Tankers and Terminals – ISGOTT (ICS, OCIMF, IAPH)*
Reference 2	*Mooring Equipment Guidelines (OCIMF)*
Reference 3	*Ship to Ship Transfer Guide – Petroleum (ICS, OCIMF)*
Reference 4	*Tandem Mooring and Offloading Guidelines for Conventional Tankers at F(P)SO Facilities (OCIMF)*
Reference 5	*Guide to Manufacturing and Purchasing Hoses for Offshore Moorings (OCIMF)*
Reference 6	*Single Point Mooring Maintenance and Operations Guide (OCIMF)*
Reference 7	*Recommendations for Equipment Employed in the Mooring of Ships at Single Point Moorings (OCIMF)*
Reference 8	*Guidelines for the Purchasing and Testing of SPM Hawsers (OCIMF)*
Reference 9	*Recommendations for Oil Tanker Manifolds and Associated Equipment (OCIMF)*
Reference 10	*Recommendations for Manifolds of Refrigerated Liquefied Gas Carriers for Cargoes 0°C to minus 104°C 2nd Ed (OCIMF/SIGTTO)*
Reference 11	*Competence Assurance Guidelines for F(P)SO Personnel (OCIMF)*
Reference 12	*Guide to Helicopter/Ship Operations (ICS)*
Reference 13	*Marine Terminal Baseline Criteria and Assessment Questionnaire (OCIMF)*
Reference 14	*ESD Arrangements and Linked Ship/Shore Systems for Liquefied Gas Carriers (SIGTTO)*
Reference 15	*International Ship and Port Facility Security Code – ISPS Code (IMO)*
Reference 16	*Guidelines for Managing Marine Risks Associated with FPSOs (OGP)*
Reference 17	*BS EN 1762 Rubber hoses and hose assemblies for liquefied petroleum gas, LPG (liquid or gaseous phase), and natural gas up to 25 bar (2.5 MPa). Specification*
Reference 18	*BS 1435-2 Rubber hose assemblies for oil suction and discharge services. Recommendations for storage, testing and use.*
Reference 19	*BS ISO 17357 Ships and marine technology - High-pressure floating pneumatic rubber fenders*
Reference 20	*IMCA M 202 Transfer of Personnel to and from Offshore Vessels.*

The Safe Transfer of Liquefied Gas in an Offshore Environment

Regulatory Compliance

1.1 General

The rules and regulations for the design and operation of F(P)SOs are not as clearly defined as those for conventional marine assets. F(P)SO technology may be considered a composite of offshore and international marine technology and practice and this is reflected in their design and operation. The particular operational risks and challenges may vary significantly from field to field and, together with Coastal State legislative requirements, must be taken into consideration in the design and operation of the unit.

The type of Classification and regulatory regime that is adopted may depend upon:

- Intended field life
- system of mooring
- whether or not the F(P)SO is to be detachable (as may be necessary in areas susceptible to hurricane activity or other seasonal changes, e.g. ice formation).

It may be necessary to assess the relevance of type of Class and regulatory approach on a case by case basis at the design stage. Class, Flag State Administration, designers, constructors, operators, oil companies and Coastal Administrations, will all have an input. The Basis of Design (BOD) of the F(P)SO will incorporate the consensus of the interested parties regarding regulatory requirements, and any deviations from typical arrangements and certification.

During conversion or construction, Classification Society surveyors will attend an F(P)SO to achieve Class certification or verification of standards applied. Where Class has been applied during conversion and mobilisation, Class may not necessarily be continued subsequent to installation and during operation. Where Class is not continued other standards should be applied. Many individual elements of the tanker rule framework may be employed. Some tanker statutory regulations are typically used, but invoking an F(P)SO interpretation. Only parts of certain IMO Resolutions may be applied.

F(P)SOs will have an interface and demarcation line drawn between the offshore design code for the topsides and the marine code areas. The demarcation will be interpreted on a case by case basis and by agreement with Classification Society and Flag State (as applicable). During the mobilisation voyage to site, marine certification (including some dispensations from normal ship regulations) may be necessary.

The rules and regulatory regime applied to F(P)SOs take into account that:

- The F(P)SO is moored at a particular location and will be marked on nautical charts as a permanent installation
- a large number of personnel onboard may be non-mariners
- the F(P)SO may or may not have the facility to disconnect and may be self-propelled
- the F(P)SO, although physically similar to a tanker, performs the function of a static terminal
- the hull is engineered for a particular site and service life, possibly without recourse to dry-docking.

1.2 Coastal State/Local Requirements

Coastal States or Governments have jurisdiction over the mineral rights within their territorial waters and may impose a variety of requirements on an F(P)SO. Rules and regulations imposed may vary from particular legislation governing the exploitation of natural resources to the Health, Safety and Environmental requirements that may influence issues, such as:

- Auditing and inspection regime
- working and living standards
- nationality, qualification, competency and rotation of crews
- control of discharged gases and waters

- noise levels
- the abandonment of the field.

National legislation may include the adoption and mandatory compliance of ISO (International Standards Organization) standards.

Depending upon the categorisation of an F(P)SO on site, i.e. a Government may not consider it a marine vessel, certain accepted marine practices and equipment may not meet the standards required by HSE regulations as applied to non-floating offshore installations and this must be considered at the design stage.

The production and export of petroleum gases will be required to meet the fiscal requirements of the governing state.

A Coastal State may impose an all-encompassing licence to operate, subject to periodic audit and verification of mandatory requirements across both the marine and production aspects of operation. In certain waters the F(P)SO is only deployed in the field offshore following government approval of a full Safety Case.

1.3 Flag State Requirements

Notwithstanding Coastal State requirements, which may take precedence, the general guidance on requirements for F(P)SO Flagging and Classification is as follows:

	Class	**Flag**
Self-propelled to site	Required	Required
Towed to site	Optional	Optional
Fixed on location	Optional	Optional
Disconnectable at site	Required	Required

Statutory requirements of the Flag State Administration may cover specific rules and regulations that an F(P)SO must meet to satisfy the Flag State Administration in a similar fashion to ships. These requirements are generally the requirements of the International Maritime Organization (IMO), embodied in such instruments as SOLAS (the International Convention for the Safety of Life at Sea) and MARPOL (the International Convention on Prevention of Pollution of the Sea) and other Conventions, enacted through national (Flag State) legislation into Merchant Shipping Law applicable to vessels under that Flag.

1.4 Classification Society Requirements

Although not strictly necessary, Classification Society rules and regulations are normally applied for assurance on F(P)SO installations and it is recommended to involve Class in the development of an F(P)SO. Classification may cover design, construction and through-life inspection and certification of a vessel (and other marine systems such as subsea moorings) based upon rules developed through experience and subject to ongoing review and revision. Members of IACS (International Association of Classification Societies), are broadly aligned on their established standards and the objective of Classification is to provide sound requirements for the construction of vessels suitable for purpose, with clearly defined limitations in the prevailing operating conditions in their locations. It is recommended the same Classification Society is utilised for classing the hull and mooring and the Certification of the process facilities during the construction and/or conversion/ integration of the F(P)SO.

En route to site, the F(P)SO may use her own propulsion systems and will, therefore, be classed for a period, after which time the machinery may be decommissioned and the Class certificates amended accordingly.

Typically, the hull is built or converted under Class and may remain in Class after installation. However, the topsides may not be classed.

Classification Societies may also be charged with responsibility for verification of F(P)SO compliance with statutory obligations on behalf of Flag and/or Coastal States. They may also verify the necessity for dispensations from certain regulatory requirements where, due to the nature of the F(P)SO, it cannot comply with specific statutory provisions.

1.5 Other Industry Guidance

Other marine industry guidelines, such as those published by OCIMF, SIGTTO, OGP and ICS, may also be applied to F(P)SOs.

The Safe Transfer of Liquefied Gas in an Offshore Environment

Safety in Design

2.1 General

These guidelines specifically address side-by-side (SBS) transfer of LPG.

Any differences applicable to LPG transfers via tandem and single point mooring arrangements are highlighted, and are summarised in Section 9. Tandem mooring operations are addressed in detail within the OCIMF publication 'Tandem Mooring Guidelines for Conventional Tankers at F(P)SO Facilities' (Reference 4).

It is important that owners, operators and project managers, responsible for developing and operating facilities for the transfer of LPG in the offshore environment, fully understand the safe working limits for the proposed operation. To properly assess the design and related operational limits, it is necessary to collect relevant environmental data and to conduct model tests based on this data for the design concept selected.

It is important that experienced operational personnel are used to assist in developing the designed operational parameters. They should also be used to assist in evaluating the results from the model testing and analysis to ensure that correct assumptions are made.

2.2 Risk Analysis

The guidance contained in the OGP publication 'Guidelines for Managing Marine Risks Associated with FPSOs' (Reference 16) should be referred to when developing a risk analysis relevant to the side-by-side transfer of LPG from an F(P)SO. The analysis should consider issues that include the following:

- Risk evaluation and management process
- design considerations impacted by
  - regulatory regimes
  - metocean specifics
  - field layout
  - field hydrocarbon characteristics
  - turret, riser and mooring systems
  - in-service life
  - hull criteria
  - major accident events (e.g. collision)
  - offloading equipment specifics
  - logistics specifics
- operational considerations impacted by
  - relationship between F(P)SO storage capacity and design parcel size(s)
  - cargo loading plans
  - cargo discharge plan and custody transfer
  - management of facility stability and stress
- specific offloading arrangements impacted by
  - major accident events (e.g. gas tanker grounding, collision, fire & explosion, pollution, export tanker breakout)
  - environmental conditions
  - emergency cargo shutdown and emergency mooring release
- communications.

2.3 Operating Environments

During the planning and development phase for any new F(P)SO development, project personnel should ensure that adequate, specific environmental data from the proposed operational location is obtained covering a sufficient time period, e.g. across all seasonal variations and preferably over several years. The ideal location for the F(P)SO should be considered at the project concept stage and should only be confirmed following review and analysis of the environmental data.

Use of long-term hindcast data is recommended to capture the extreme conditions to be taken into account in the facility's design. The environmental data should be used in model tests and simulations to determine the optimal mode of operation, mooring equipment layout and predicted loads, fendering requirements, the reaction of different size design gas tankers, and to ensure that predicted downtime due to environmental conditions is within the specified requirements for the project.

Site specific information requires to be collated as follows:

- Wind speed and direction
- wave height and period (maximum and significant)
- swell height, direction and period
- current data, direction and rate at various depths
- the duration of metocean conditions above a threshold value that would affect berthing and transfer operations
- tidal variation
- air temperatures
- sea temperatures
- precipitation rates and type
- wind squall activities and other extraordinary events, e.g. soliton
- visibility that may be restricted due to local phenomena, such as Harmattan in West Africa
- ice formation.

It is important that all data above is correlated to ensure that combined effects are understood.

2.4 Model Tests and Simulations

Model tests are undertaken using scale models in a purpose built test tank that has the capability to replicate the anticipated environmental conditions at the proposed location. Simulations are computer-based assessments of F(P)SO and vessel motions and interactions in differing environmental conditions and are of particular benefit when, for example, evaluating manoeuvring and berthing operations or the need for tug or thrusters support. When evaluating a new project proposal it is recommended that both model tests and simulations are undertaken.

It is recommended that model tests and simulations are used to determine the environmental (e.g. wind, current and sea state) and operational limits and requirements for activities that may include:

- Personnel and equipment transfer
- launching and retrieval of fenders
- approach and berthing, including the use of tugs number, type, power and location requirements and the identification of manoeuvring and restricted areas
- station keeping of the F(P)SO and the maintenance of directional stability and the use, size and location of thrusters

- mooring
- handling, connection and disconnection of transfer hoses/arms and the dynamic forces imposed on them
- cargo transfer
- unmooring, including emergency departure of the gas tanker.

Figure 1: Model Test

The model tests and simulations should take into account the full range of gas tanker types and sizes that it is proposed are to be handled at the facility.

The output from the modelling and berthing simulations should be validated by an experienced Mooring Master to ensure that limits are properly assessed. This will also enable Mooring Masters to develop contingency scenarios for situations such as those involving berthing, mooring, unmooring and tug or thruster failure.

The berthing simulation output should also be incorporated into ongoing training and assessment tools for the Mooring Masters that will be conducting berthing operations at the facility.

It is important that the simulation process includes the capabilities of tugs and support craft so that their mode, placement and power can be properly assessed. It should be demonstrated that line handling boats are fit for purpose, including their ability to be operated, launched and recovered in the limiting environmental conditions established for the location.

2.5 Tug Support

A variety of factors must be considered to establish how many tugs are required to assist in safely berthing a gas tanker to an F(P)SO, including:

- Whether the F(P)SO is turret moored or spread moored
- environmental conditions at the location
- experience of the terminal operators and Mooring Masters
- sizes and manoeuvrability of gas tankers to be handled.

As a general rule, a weather vaning F(P)SO will require less tug assistance than a spread moored F(P)SO.

Tugs used to assist the gas tanker during berthing and un-berthing in an SBS operation may be considered in two categories:

- Those which work alongside the gas tanker, i.e. in push/pull mode
- those which operate on a long tow line.

Tugs Alongside
The environmental limitations for tugs working alongside vessels in the open water locations of F(P)SO facilities will dictate the operable limits for mooring gas tankers. Tug motions alongside may potentially lead to damage to the gas tanker and/or tug. Additionally, tug movement impairs the ability of the tug crew to safely handle towing lines. Alongside work in open sea conditions should only be performed by tugs that are designed and strengthened for the purpose, and within strict operating limits.

When considering suitable tugs for working alongside the gas tanker, the following should be considered:

- Number of tugs required
- fendering arrangements (on pushing point and side of tug)
- effectiveness of tug in pushing mode
- power and bollard pull
- propulsion arrangements
- design of towing winch, including consideration of auto tension winches
- tow line material
- adequate freeboard to protect personnel and equipment during operations.

Tugs on a Long Line
Tugs of adequate capability should be provided for berthing/unberthing assistance and to maintain the gas tanker at a safe distance from, and in correct alignment with, the F(P)SO during berthing operations. The tow should be capable of being maintained in all weather conditions up to the maximum operable conditions for the F(P)SO SBS operation. As the environmental conditions increase tug efficiency will be progressively restricted, and tug effectiveness may become a limiting condition requiring operations to stop.

The following points should be considered:

- Number of tugs required
- power and bollard pull determined from a mooring design/simulation study
- manoeuvrability and propulsion arrangements
- ability to tow over the bow and/or stern
- towing equipment.

Failure to select the most suitable tug may result in an increased risk of an incident between the F(P)SO and the gas tanker or reduced operational availability of the terminal.

2.6 Berthing Aids

The terminal design will have evaluated angles and speed of approach of gas tankers and have determined the recommended angles and preferred speed. To effectively monitor the gas tanker approach to the F(P)SO, consideration should be given to utilising berthing aids, such as speed of approach monitors, to minimise the risk of damage to either the F(P)SO or gas tanker.

As a minimum, the berthing aid used should be capable of displaying the following information:

- F(P)SO heading
- gas tanker heading
- relative approach angles
- rate of turn of F(P)SO
- rate of turn of gas tanker
- approach speed of gas tanker
- relative distance from bow and stern of the gas tanker to the F(P)SO
- prevailing wind speed and direction.

Additionally, information on prevailing current strength and direction should be made available.

Established technology employed in berthing aids includes laser, GPS, DGPS, or a combination of these.

The terminal should have a process in place to ensure that berthing aids are properly maintained and kept operational.

It is recommended that the information from the berthing aids is available on the bridge of the approaching gas tanker.

2.7 F(P)SO Mooring System

There are a number of mooring systems employed for F(P)SOs. However, they largely fit into two separate categories, weather vaning or fixed heading (spread moored).

In general, SBS operations would not normally be conducted at a spread moored facility unless it is in a very benign weather and current area, or where the heading of the facility can be aligned to the predominant current flow, such as in a river or channel.

Where a facility is to use SBS for offloading in an open sea environment, it is recommended that a weather vaning F(P)SO is utilised as this gives greater control over the offtake operations and allows the F(P)SO to head in the most appropriate direction for berthing and offtake operations.

Where an F(P)SO is spread moored in an offshore environment, it may be more appropriate to use either a tandem mooring arrangement for the gas tanker or a remote offloading system, such as an SPM, or a novel system such as a dynamic position loading terminal.

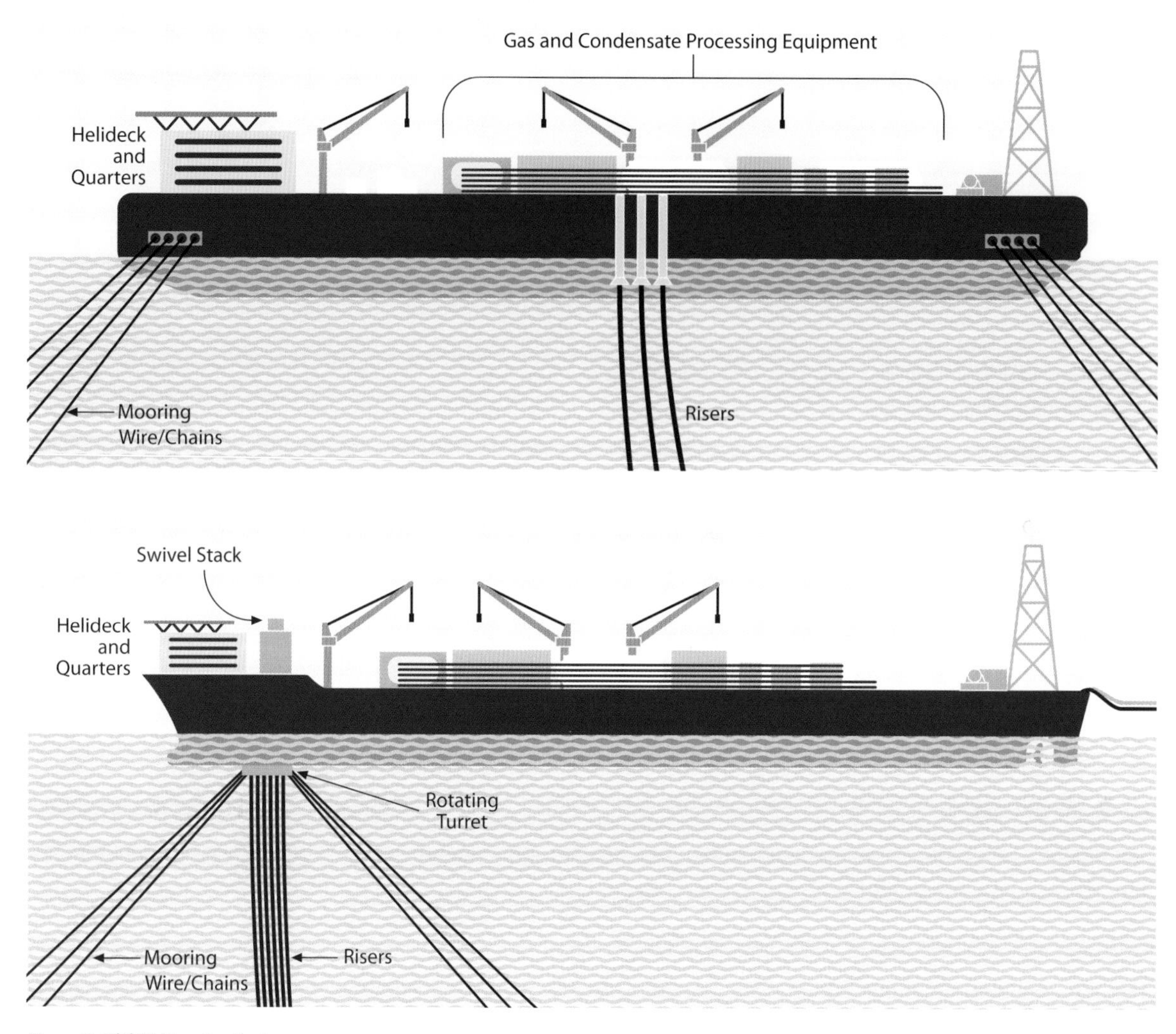

Figure 2: F(P)SO Mooring Systems

2.8 F(P)SO Heading Control

Where an F(P)SO is weather vaning and SBS is the method to be used for the cargo transfer operations, effective heading control of the F(P)SO is important to ensure a safe berthing operation and to maintain the integrity of the moorings between the gas tanker and the F(P)SO.

Generally, there are two methods for maintaining heading control of the F(P)SO during the berthing and cargo transfer operations:

- Fitting a thruster to the F(P)SO that is capable of maintaining the desired heading associated with the operational limits for the operation.
- use of tug(s) having sufficient power and manoeuvrability to maintain the desired heading associated with the operational limits for the operation.

Sizing of either the thruster or tug(s) should be determined during the simulations and be aligned to the range of predicted environmental conditions for operations, including any project-specific redundancy requirements.

Operational experience has shown that it is critical to avoid excessive rolling of the F(P)SO and/or the gas tanker while moored alongside. This is to avoid snatch loading on the mooring lines, which may cause failure and breakout of the gas tanker from the F(P)SO. For this reason, heading control must be maintained throughout the SBS operations to minimise rolling. If it is not possible to maintain heading control via either a thruster or tug(s), consideration should be given to stopping the operation and unmooring the gas tanker from the F(P)SO. The gas tanker should be moved clear to a safe distance until heading control can be re-established.

Mooring line load tensions are a good guide to the most appropriate heading for the cargo transfer operations. Line loads should be kept to as close to a constant value as possible, without sudden decreases and increases in load that would indicate snatching is occurring.

Heading control of the F(P)SO may also be useful in achieving a lee for the launching and recovery of fenders and for the transfer of personnel and equipment.

Where a thruster is to be used for heading control, designers have two types to choose from:

- Tunnel thruster
- azimuth thruster.

Both types provide an acceptable means of achieving heading control, but interaction of the thruster on the hull of the gas tanker during Side-by-side operations should be considered when deciding which type to install. The azimuth thruster is mounted below the hull of the F(P)SO and should interact less with the hull of the gas tanker than a tunnel thruster. Consideration should also be given to the associated in-service maintenance requirements and methods.

Where thrusters are used, it is recommended that they are linked to an automatic active heading control system.

2.9 Mooring Equipment Layout Configuration and Compatibility

The ability to safely moor and unmoor the gas tanker to the F(P)SO, and to maintain the gas tanker's mooring integrity throughout gas transfer operations, is critical for safe operations.

The following are among the mooring equipment issues that should be determined early in the project design stage:

- Gas tanker and F(P)SO dimensions, including freeboard range
- range of gas tanker manifold offsets and the impact on the mooring arrangements
- range of gas tanker types (e.g. fully refrigerated or fully pressurised)
- cargo transfer systems to be employed
- gas tanker manifold arrangements (refer to OCIMF/SIGTTO's 'Recommendations for Manifolds of Refrigerated Liquefied Gas Carriers for Cargoes 0°C to minus 104°C 2nd Ed' (Reference 10)).

From the results of the model tests (Section 2.4), and taking into account the above factors, the expected mooring line loads, at each individual mooring hook/bitt on the F(P)SO, for the range of gas tanker sizes to be moored alongside the F(P)SO for the full range of metocean scenarios, should be determined.

Typical general arrangements for a sample of vessels, that includes the smallest and largest gas tankers (to the same scale as the F(P)SO's general arrangement), should be obtained and used to identify the various optimum mooring arrangements and relative positions of gas tanker/F(P)SO location for each size of gas tanker.

It is recommended that a remote release capability is provided for all moorings between the F(P)SO and the gas tanker in order that personnel are not put in danger should there be a need to release the moorings in an emergency. It is recommended that quick release hooks (QRHs) are fitted to the F(P)SO, with the remote release capability being initiated from the F(P)SO control room. In designing the remote release system, it is recommended that single activation release of individual lines is preferred over a gang release type of arrangement, which may lead to the inadvertent release of all moorings. Single activation of the hooks allows a controlled approach to the release of moorings. The use of portable QRHs rigged on the gas tanker is generally not recommended, but is preferable to having no quick release capability.

The QRHs located on the F(P)SO can be placed appropriately to cover the variety of gas tankers expected. If using the larger vessels as an example, OCIMF's 'Ship to Ship Transfer Guide – Petroleum' (Reference 3), recommends a minimum of 14 lines be run between the vessels. If an F(P)SO is expected to handle this size of vessel, experience indicates that at least 20 QRHs will be needed to cover the wide range of vessels likely to berth at the facility.

The QRHs and associated closed fairleads should be distributed on the F(P)SO so as to give the best possible line leads. Lines between the gas tanker and the F(P)SO should be at the lowest angle possible and also have as long a lead as possible without interfering with manifold areas or fendering arrangements.

Figure 3: View of Side-By-Side Transfer Operation

Once the optimum mooring layout has been determined, it is recommended that the arrangement is confirmed by re-assessing the individual maximum mooring line loads, either via simulation or model testing, to ensure that the layout will perform as expected. If it is planned to handle gas tankers that are of a similar length to the F(P)SO, particular attention will need to be given to the potential for increased snatch loading on lines due to the short run of the moorings between the F(P)SO and gas tanker. The freeboard differences between the two vessels will also need to be considered as this will impact on the holding force applied by the moorings and on their effective safe working loads.

In general, it is preferred that mooring lines are run from the gas tanker to the F(P)SO in order that the quick release capability is maintained. Where there is a need to run lines from the F(P)SO to the gas tanker to make up the required numbers, it is preferable to use an arrangement that will also maintain overall quick release capability, such as one using grommets (see Section 2.15.3). Any arrangement should be thoroughly analysed and risk assessed before being adopted, to ensure that it will meet the operational requirements.

Conventional synthetic fibre mooring lines and steel wire ropes should not be used in the same service owing to their differing elasticity. Synthetic fibre ropes manufactured from high modulus synthetic fibres, such as Aramid or HMPE, are acceptable for service in conjunction with wire moorings as their elasticity is similar to that of wire.

Due to the dynamic nature of SBS operations it is important that line integrity is maintained. The F(P)SO should be fitted with enclosed style fairleads of appropriate capacity that may be of the panama or roller type in way of all QRHs and winches. If panama style fairleads are used, they should be maintained in a smooth (polished) condition to prevent chaffing of synthetic tails fitted to wires or, if used, high modulus mooring lines.

Gas tankers must be fitted with enclosed fairleads of either panama or roller construction. If roller fairleads are used, they must be of the enclosed type.

All mooring lines, whether wire or high modulus synthetic fibre, should be provided with a suitable synthetic tail of at least 11 metres in length. It may, however, be configured as a grommet, as noted on page 16, on lines that are deployed from the F(P)SO. For SBS operations it has been found that longer tails of up to 22 metres in length for headlines provide better dampening than the shorter tails and prevent early line failure. As gas tankers do not normally carry tails of this length, it is recommended that the F(P)SO operator carries a sufficient stock so that the gas tanker's tails can be changed before mooring operations commence.

Another design consideration relates to the SWL of the large number of individual components creating the overall mooring arrangement. The mooring arrangement can include the F(P)SO and gas tanker components including, but not limited to the following, the SWL of which should be confirmed as compatible with the modelled/simulated mooring line loads:

- Mooring winches
- mooring lines
- mooring tails and connecting shackles
- F(P)SO QRHs
- mooring bitts
- F(P)SO 'grommet' arrangements

Reference should be made to 'Mooring Equipment Guidelines' (Reference 2) for detailed information on the design requirements and safety factors for mooring fittings.

The final range of gas tanker mooring arrangements required to ensure the maintenance of the mooring integrity of the gas tanker to the F(P)SO should then be used to:

- Identify potential manifold arrangement offset issues (see Section 2.11)
- identify issues associated with the mooring and unmooring operations.

To identify issues associated with the gas tanker mooring and unmooring operations, an operational hazard identification study (HAZID) should be conducted to determine and validate the appropriate mooring philosophy to be employed and to validate the final design assumptions. This HAZID should also pay particular attention to emergency situations that may affect both the F(P)SO and the gas tanker, e.g. uncontrolled gas release, fire and gas tanker break-out due to mooring failure. In conducting the HAZID it is important to involve personnel that have appropriate operational experience.

It is recommended that all line handling undertaken on the F(P)SO uses winches rather than capstans. This allows for the 'hands free' retrieval of mooring lines and minimises the risk of injury to line handlers. Ideally, messenger lines from the gas tanker should be led directly to winch storage drums on the F(P)SO and the arrangement should result in the mooring line's eye running to a position over the appropriate QRH. The messenger line can then be slacked back in a controlled manner so that line handlers only need to guide the eye of the mooring line onto the QRH, rather than having to man-handle the eye onto the hook.

Consideration should be given to the use of high modulus synthetic fibre mooring lines rather than wires as they are easier to handle, lighter, do not pose injury risk through broken wires and can be used with the F(P)SO 'grommet' stretcher arrangement.

When mooring the gas tanker to the F(P)SO, the mooring line transfer procedure should consist of all mooring lines being initially transferred from the gas tanker to the F(P)SO. Any supplemental mooring lines from the F(P)SO should then be transferred. Given this, together with the general requirement to commence transferring the gas tanker's mooring lines at the earliest opportunity, the mooring lines can be transferred from the gas tanker to the F(P)SO using the existing transfer methods, such as a line handling boat. In the absence of a line handling boat, the use of a non-hazardous, pneumatic line throwing apparatus should be considered as a potential alternative. However, such a device should only be considered for use onboard the F(P)SO by appropriately trained personnel, and not onboard the gas tanker.

Pneumatic line throwers may require use of an additional, smaller diameter messenger line originating from the F(P)SO, which, deployed by the throwing apparatus line, can be used to recover the gas tanker's messenger line for transfer of the gas tanker's mooring lines.

As the separation distance between the gas tanker and the F(P)SO reduces, traditional heaving lines may be used to transfer lines.

When unmooring the gas tanker from the F(P)SO, the design of the mooring arrangement should not hinder the overall operation. Any supplemental mooring lines from the F(P)SO should be capable of being released first, followed by the gas tanker's mooring lines.

Figure 4: Pneumatic Line Thrower (PLT)

2.10 Fendering Equipment

To safely conduct SBS operations, it is important that the correct fenders are used to ensure that the gas tanker and F(P)SO are kept far enough apart that they do not contact each other, and that suitable energy absorption exists to prevent hull plate damage to either vessel.

As the F(P)SO will always be the mother vessel, it is recommended that the fenders are deployed from the F(P)SO.

The use of fixed, jetty style, fendering is not recommended for offshore SBS operations for a number of reasons:

- The amount of internal structural steel required to support this style of fendering is significant and may need to extend into the cargo tank area
- structural fatigue will be greatly increased around the area of the fendering due to the loads applied
- the fenders are fixed and not capable of being repositioned to meet different gas tanker hull shapes and lengths
- repairs require over-side work, which increases the risk to personnel
- mooring lines are more easily fouled on this style of fender.

In general there are two categories of fenders used for SBS operations offshore, primary and secondary fenders, and these are described in the following sections.

2.10.1 Primary Fenders

These are fenders that are positioned along the side of the F(P)SO where the parallel body of the gas tanker will rest during the cargo transfer operations to afford maximum possible protection while alongside. Primary fenders may be foam filled or pneumatic type (0.5 to 0.8 kg/cm^2) and should be manufactured, tested and maintained in accordance with industry and international standards. An International Standard, ISO 17357 (Reference 19), specifies the material, performance and dimensions of floating pneumatic fenders and it is recommended that any pneumatic fenders used for SBS operations comply with this or an equivalent standard.

The fenders may be secured on either the F(P)SO or the gas tanker, but it is recommended that they are secured to the F(P)SO as this will limit the amount of manual handling and reduce the risk to personnel. It is also recommended that primary fenders be deployed and retrieved by purpose built davits rather than by crane or boat as this again reduces the risk of manual handling injuries to personnel. It is recommended that falls are removed from the fenders once they are deployed unless the arrangement includes an adequate load compensation system to prevent snatch loading and/or entanglement of the falls.

Figure 5: Davit Launched Primary Fenders

The sizing of the fenders needs to take into account:

- The size of the gas tankers to be moored alongside
- the weather conditions anticipated during the operation
- the relative freeboards of the gas tanker and the F(P)SO during the transfer operation
- the effect asynchronous rolling has on the two bodies and the separation distance required.

Reference should be made to the guidance on fender selection contained in the ICS/OCIMF publication 'Ship to Ship Transfer Guide – Petroleum' (Reference 3), and that provided by individual fender manufacturers, when determining the number and sizes of fenders for a particular operation

The primary fenders should be secured alongside via a dual mooring arrangement to give redundancy in case a mooring line breaks. It is usual that fenders will be rigged in pairs with a mooring line between the two fenders in the pair. The location of the primary fenders should be clear of hose deployment and manifold areas.

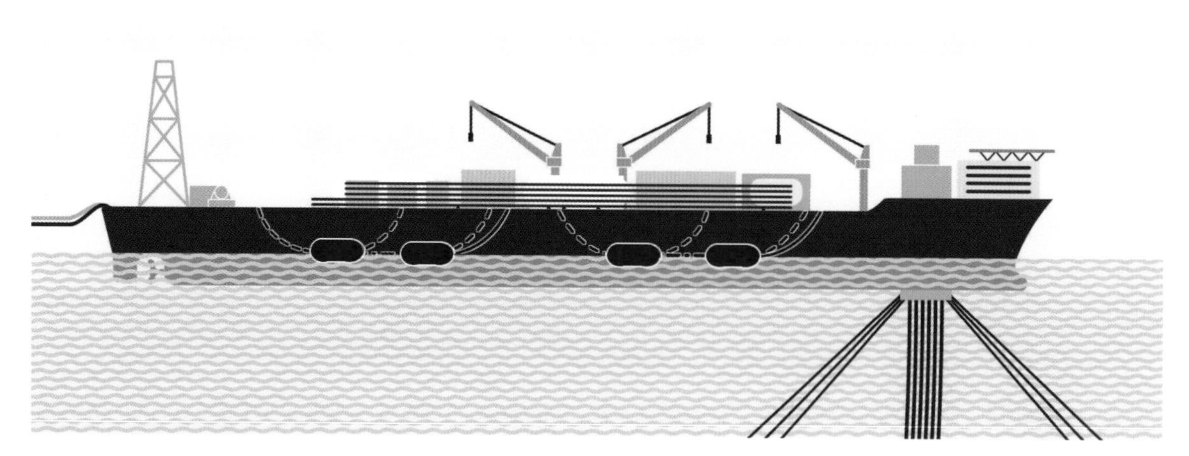

Figure 6: Fenders Rigged on an F(P)SO

2.10.2 Secondary Fenders

Secondary fenders are used to protect the bow and stern plating from inadvertent contact during berthing and unberthing operations if the gas tanker and F(P)SO get out of alignment. The point(s) where such contact is likely to occur may be high up due to the flare of the bow or stern. It is recommended that either enough fenders are rigged on the F(P)SO to cover change in height requirements, or that the fenders are able to be adjusted during the berthing or unberthing as their height requirements change. Secondary fenders may be foam filled or pneumatic type. They are usually much smaller than the primary fenders and, if they are to be moved, it is advantageous for them to be of a lightweight construction.

Consideration should be given on a case-by-case basis to any potential benefits of rigging secondary fenders on the gas tanker.

2.11 Cargo Transfer Equipment

2.11.1 Hoses

The transfer of LPG in an offshore SBS operation is usually undertaken by the use of hose(s). The three types of cargo hose suitable for the transfer of liquefied gases are of composite, rubber or stainless steel construction.

2.11.1.1 Hose Standards

It is important that transfer hoses used are specially designed, constructed, tested and rated for the LPG product to be handled and the expected surge and locked in pressures associated with the product. Hoses should comply with BS EN1762 (Reference 17) or an equivalent standard with regard to the assemblies. When purchasing hoses, the important specifications to be relayed to manufacturers include:

- The maximum and minimum operating temperatures and pressures
- the compatibility of the hose material with the cargo
- the minimum bending radius.

As hoses are to be handled from the F(P)SO to the gas tanker, it is important that they are as flexible as possible to allow easy connection to the gas tanker's manifold.

2.11.1.2 Hose Size and Length

It is important that the transfer hoses are sized appropriately for the designed flow rate and are not oversized as this only makes handling and connection more difficult. Hoses in excess of 300 mm (12 inches) in diameter will be progressively more difficult to handle and particular care will be needed to avoid damage from kinking unless the hose assembly is specifically designed to overcome this problem.

Hose(s) should be of suitable length so that they can be easily handled by the operations personnel, can allow for the offset of the F(P)SO and gas tanker manifolds and allow for changes in relative freeboard between the gas tanker and the F(P)SO during the operation and prevent kinking.

2.11.1.3 *Hose System Fittings*

Hoses should be capable of being isolated by means of a valve on the tanker rail end. A hose end valve should be incorporated at the end of the tanker rail hose with a suitable spool piece to connect to the gas tanker's manifold. It is recommended to use a lockable wafer-type butterfly valve without tapped lugs in the hose end, as the loads are taken through the bolts that capture the valve body between the hose flange and end spool piece, with the valve itself isolated from the hose lifting loads.

To protect against the risk of electrical arcing at the manifold during connection and disconnection of cargo hoses due to the possible differences in electrical potential between the F(P)SO and the gas tanker, a means of electrical isolation should be included in the hose assembly. The electrical isolation should be provided by an insulating flange fitted on the F(P)SO close to the presentation flange. The insulating flange's position in the pipe work should ensure that no supports to the F(P)SO deck exist between the insulating flange and the point of hose connection on the gas tanker. As an alternative to an insulating flange, one length only of electrically discontinuous hose may be included in each hose string.

In designing the hose system, consideration should also be given to the method of inerting hoses prior to their disconnection and storage. If nitrogen from the F(P)SO is to be used, it is recommended that a connection for a small diameter hose between the F(P)SO's nitrogen delivery outlet and the gas tanker's manifold drain valve is provided, allowing liquid and gas in the hose to be pushed back to the F(P)SO.

2.11.1.4 *Hose Handling*

If the arrangement for the transfer and connection of hoses is designed to make use of the gas tanker's lifting gear, due consideration must be given to the safe working load (SWL) in relation to the anticipated weight to be lifted.

If the design makes use of the F(P)SO's lifting gear, consideration should be given to locating the controls for the crane as close to the side as possible to provide the operator with a clear view of the gas tanker's manifold area.

It is recommended that the hose deployment method should be such that, once the hoses are attached to the gas tanker's manifold, neither the F(P)SO crane(s) or the gas tanker's lifting gear remain attached to the hose. If cranes or derricks are left attached there is a potential for serious damage to the equipment in an emergency release situation should the hoses not release in time. Should an operator wish to have the crane remaining attached to the hose during an SBS operation, it should be ensured that the method is properly risk assessed and that appropriate measures are implemented to prevent damage to the crane in an emergency release situation.

Figure 7: Transfer Hoses Unsupported by Crane

Figure 8: Transfer Hoses Supported by Crane

2.11.1.5 Hose Storage

It is important that storage arrangements are included in the design stage for the hose system, including provision for routine inspection and testing. When hoses are being stored between transfers it is recommended that they either be removed from the F(P)SO manifold and laid out flat, or be hung along the side with an appropriate rack for the gas tanker's manifold hose end. If the hoses are to be stored alongside, the rack should be sufficient distance away from the F(P)SO hose end so that the hose is not forced into too tight a bend radius, which might cause damage

It should be determined whether the hoses will be left gas (not liquid) filled or inerted in between offtakes. If hoses are to be disconnected and laid flat, it is recommended that they are inerted before disconnection. If the hoses are to be stowed alongside the F(P)SO, it is recommended that they remain in a gas filled (liquid free) state to avoid the gas tanker having to handle inert gas at commencement of cargo transfer.

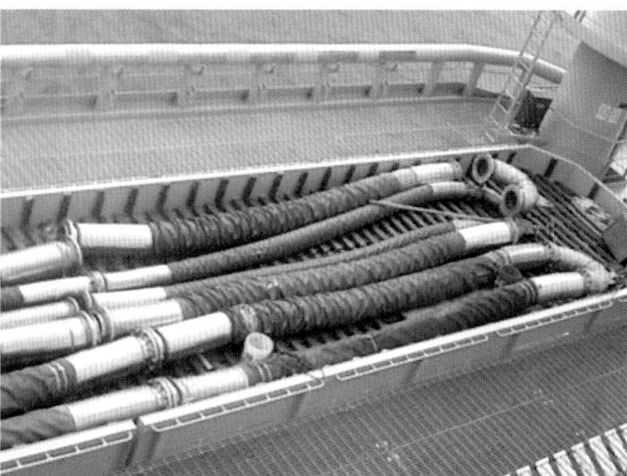

Figure 9: Examples of Hose Storage Arrangements

2.11.1.6 Hose Inspection and Testing

Hoses should be subjected to regular in-service inspection for damage or deterioration. A record of inspection and pressure/vacuum testing should be maintained.

Periodic testing of hoses should be in accordance with the requirements of the specification to which the hose was manufactured and/or as detailed in accordance with British Standards BS1435-2 (Reference 18) as amended, or equivalent.

In consultation with the hose manufacturer, the retirement age for the hoses should be defined to determine when they should be removed from service.

2.11.2 Hard Arms

Hard arms are not commonly employed in SBS operations involving the transfer of LPG because of the relative vessel motions and the potential for encountering high accelerations in the offshore environment. However, their use does result in less manual handling than when using hoses and the design can incorporate an emergency release function in which the arms are released in a safe and controlled manner.

The following are amongst issues to be considered when assessing the option to use hard arms:

- The use of model tests to validate the limits for the arm's operating envelope
- the provision of position monitoring that includes range of motion and excursion alarms
- the inclusion of emergency shutdown and release functions
- methods of connection to the gas tanker's manifold based on functionality of arm disconnection
- use of combination systems, ie a hard arm provided with a length of hose, to provide greater flexibility in the system.

Hard arms should be provided with an insulating flange that is normally located in way of one of the arm's swivel joints.

2.11.3 Quick Connect/Disconnect Couplings (QC/DC)

Quick Connect/Disconnect (QC/DC) couplings are used to facilitate faster connection and disconnection of hoses to the offtake tanker manifolds than using bolted flanges. The QC/DC coupling should not be confused with Emergency Release Couplings (ERCs), which are activated in an emergency situation. A QC/DC usually has no remote activation capability.

QC/DCs are commonly available as either manual or powered activated.

2.11.3.1 *Manually Activated QC/DC*

A manually activated QC/DC usually incorporates a cam locking device that has rotating cams located around a central ring with an 'O' ring seal.

Figure 10: LPG Hose Manally Activated QC/DC with Blank

2.11.3.2 *Power Activated QC/DC*

The coupling is usually operated either electrically or hydraulically and has clamps that hold the hose onto the tanker's offloading manifold. Power activated QC/DCs are not typically used in conjunction with transfers by hose due to their weight and complexity. They are more typically associated with hard arm applications

2.11.4 Gas Tanker Manifold Arrangements

The design of the F(P)SO's cargo transfer system should take into account the recommended arrangement for gas tanker manifolds as described in the OCIMF/SIGTTO publication 'Recommendations for Manifolds of Refrigerated Liquefied Gas Carriers for Cargoes 0°C to minus 104°C 2nd Ed' (Reference 10).

In addition, the system design should consider the specific requirements for non-standard operations, such as those associated with commissioning and start-up activities.

2.12 F(P)SO Venting Systems

A gas dispersion study should be undertaken to assess the optimum position of all intakes and vents on the F(P)SO.

It is recommended that all production and cargo venting on the F(P)SO is led to a flare or is handled by a vapour recovery unit. If this is not possible, vents should be sited as far from the cargo transfer area as is practicable. The vent should not be sited above the deck as hydrocarbons emitted will fall to the deck in still conditions, creating a potentially hazardous situation.

The F(P)SO design should ensure that hydrocarbon gases are not vented to the cargo deck area.

During cargo transfer operations, gas tankers should conduct operations without venting. To cater for increased cargo transfer rates and to provide contingency in the event of reliquefaction plant failure on the refrigerated gas tanker, consideration should be given to incorporating a vapour return capability in the F(P)SO's system design.

The return of vapours to the F(P)SO may be subject to local fiscal restrictions, in which case they may require metering. In addition, there is a possibility of contamination of the F(P)SO's tanks due to the unknown composition of the vapours received from the gas tanker.

2.13 ESD Systems

The design study should take into account the requirements for emergency shut down systems. It is recommended that any system on the F(P)SO for the cargo transfer system is designed and installed in line with the recommendations contained in the SIGTTO publication 'ESD Arrangements and Linked Ship/Shore Systems for Liquefied Gas Carriers' (Reference 14).

2.14 Emergency Release Couplings

To achieve quick release of the gas tanker from the F(P)SO in an emergency, it is recommended that all hoses used in SBS LPG transfers are capable of being released remotely from the manifold station or gas tanker manifold in a safe manner that does not result in a large release of liquid or gas. This recommendation applies equally to any vapour return hoses that may be used.

The ERCs should include the provision of two ball valves that close either side of the ERC release point to prevent product release. Consideration should be given to purging the cavity between the two valves to minimise hydrocarbon release.

QC/DCs as described in Section 2.11.3 should not be considered as ERCs as they are not designed to be remotely activated and are not able to seal off both sides (i.e. manifold and hose) to minimise liquid or gas release.

The use of Marine Breakaway Couplings instead of an ERC is not recommended due to the forces that would be imposed on the gas tanker's manifold prior to activation, which could result in structural damage to the gas tanker.

2.14.1 Location of ERC

2.14.1.1 Location of ERC on Hoses Unsupported by Crane

If hoses are not being supported by crane(s), the ERC is best sited onboard the F(P)SO outboard of the hull. This configuration will allow the hose to fall free of the F(P)SO and will remain attached to the gas tanker's manifold as the vessel moves clear of the F(P)SO. The hose end that has the release coupling attached should be provided with a buoy to prevent the hose end sinking, making retrieval difficult. The coupling half that remains with the hose should also have a protection rail fitted to prevent damage to either the coupling or the F(P)SO hull should it strike the side on release. This should be constructed of a non-sparking material.

Figure 11: Emergency Release Coupling

2.14.1.2 *Location of ERC on Hoses Supported by a Crane*

If hoses are to be supported by a crane during the cargo transfer operation, the ERC is best sited on the connection between the hose and the gas tanker's manifold. This configuration will allow the hose to be held clear of the manifold as the gas tanker is moved clear of the F(P)SO. In considering this possible arrangement for positioning the ERC, the design of equipment to be transferred to the gas tanker and the associated difficulties of handling the ERC during hose connection should be assessed.

Operators should be aware that should the ERC fail to release, there is a risk that the crane supporting the hose will come under considerable strain if the two vessels are moved apart, which might result in catastrophic failure of the crane.

Figure 12: Portable Emergency Release Coupling

2.15 Quick Release Capability of Moorings

As described in Section 2.9, it is recommended that mooring system design includes the capability to remotely release moorings in a safe and controlled manner. This may best be achieved by the use of quick release hooks (QRHs).

2.15.1 Fixed Quick Release Hooks

Fixed quick release hooks are mooring hooks that are fitted to secure bed plates on the F(P)SO. The hooks themselves usually have the ability to swing through the horizontal plane and may have limited movement in the vertical plane. The actual hooks usually rotate about an axle when released, causing the jaw of the hook to fall free, allowing the eye of the line to fall clear.

Three types of remote activation methods are commonly used for releasing mooring hooks: pneumatic, hydraulic and electric. These are in addition to local manual release. Risk analysis should be used to determine the most appropriate activation method. No matter which activation method is chosen, local manual activation should be retained. The hooks should also be connected to the facility's uninterrupted power supply (UPS) system to ensure that they are able to operate in an emergency situation if power is lost.

It is recommended that mooring(s) between the gas tanker and the F(P)SO are capable of being released both remotely from the FPSO's Central Control Room and locally at the quick release hooks. It is particularly important that QRHs are able to be individually released remotely so that personnel are not put at risk if an emergency situation develops that could expose personnel on deck.

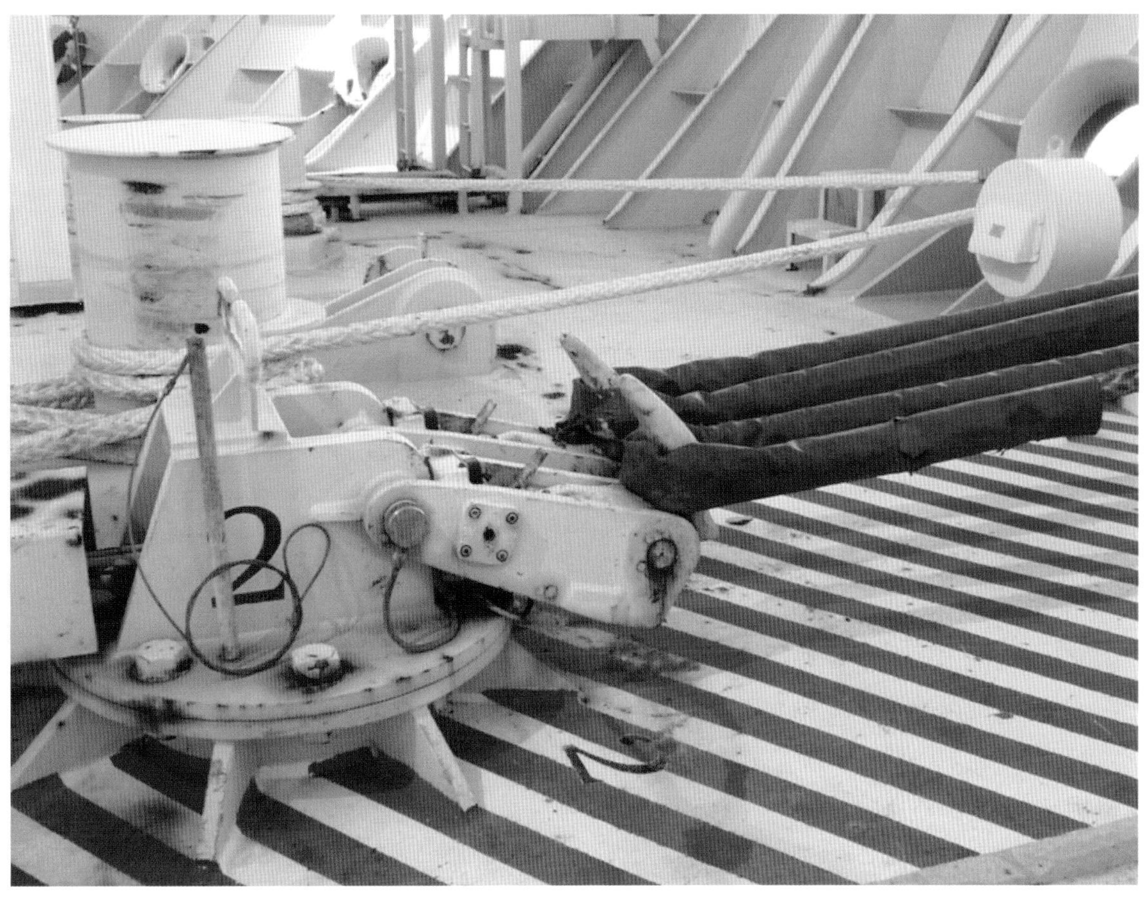

Figure 13: Fixed Quick Release Hooks

All QRHs should be fitted with load cells. It is recommended that the remote load monitoring facility is available within both the F(P)SO's CCR and to the officer on watch on the gas tanker. On the F(P)SO this will normally be accomplished via a hard wired feed to the CCR, and for the gas tanker a wireless feed is usually employed to a laptop computer carried by the Mooring Master or Assistant Mooring Master(s). The line loads should also be stored in a database or be recorded in some other manner, so that they may be assessed against the design case. This will also to assist in determining the retirement criteria for any F(P)SO lines or tails used in the mooring operation.

QRHs should be sized appropriately to the range of tankers that are expected to berth at the F(P)SO.

2.15.2 Portable Quick Release Hooks

Portable QRHs are made to be attached to the gas tanker's mooring bitts to provide a quick release capability in case of an emergency. Portable QRHs are secured to the mooring bitts by the use of a wire.

Portable QRHs need to be sized appropriately to give the same holding capacity as the fixed QRHs on the F(P)SO.

Figure 14: Portable Quick Release Hook

Most portable QRHs are manually released at the hook. For this reason they are not recommended for use in SBS operations involving gas tankers as they may pose a risk to personnel in an emergency situation that requires the gas tanker to quickly unmoor from the F(P)SO.

2.15.3 Use of Grommets

A grommet provided by the F(P)SO can supplement the gas tanker's moorings by using moorings deployed from the F(P)SO while retaining the capability to control the quick release of all moorings.

A grommet arrangement involves the use of a stretcher (synthetic tail) designed in a continuous grommet formation, with an oversized polished stainless steel thimble at one end and a joining shackle (e.g. Mandel, Tonsberg or Boss shackle) at the other. On the end with the joining shackle there should be a wire with a spliced loop at its other end. The loop should be of a size that allows it to fit over the mooring bitts of the gas tanker. The stretcher part should be at least eleven metres in length, and the wire at least six metres, to ensure that it is clear outboard of the gas tanker's fairlead. The F(P)SO should have high modulus synthetic fibre mooring lines that can be passed through the stainless steel thimble and led back to the QRH. The wire at the other end is pulled over to the gas tanker and made fast to the appropriate mooring bitts. The wire is used to prevent failure due to chafing as the fairleads used are usually panama style enclosed chocks. The grommet stretcher will remain outboard of both the F(P)SO and the gas tanker.

If an emergency disconnection is required, QRH activation enables the F(P)SO's mooring line to slip through the thimble on the grommet. In this way, remote release capability can be retained even if lines are run from the F(P)SO to the offtake tanker.

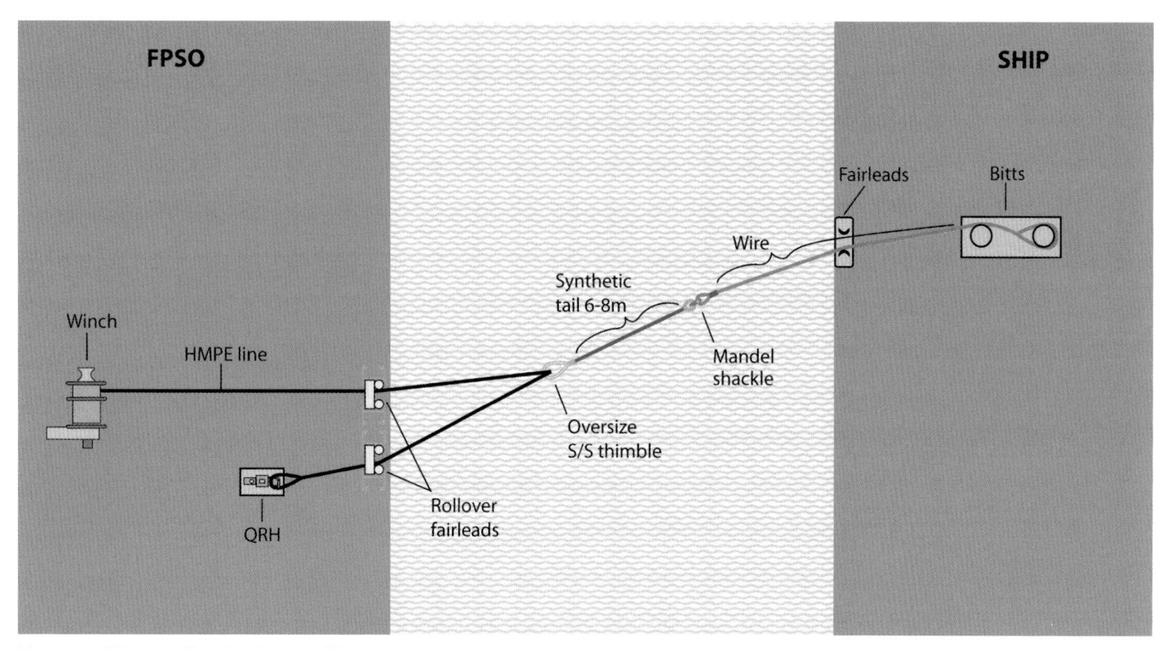

Figure 15: Diagram Showing Layout of Grommet Arrangement

Figure 16: Grommet Arrangement Showing Stretcher and F(P)SO Mooring Passed Through Large Thimble

2.15.4 Use of Tugs in an Emergency

In order for the gas tanker to be manoeuvred clear of the F(P)SO in an emergency, it is recommended that operators keep the tug(s) made fast on a line. The line length should be such that the tug will be clear of potential hazards associated with the offtake operations, such as gas venting, fire, etc. The length of line needs to be determined by the operator, but a length greater than 50 metres is recommended. If the operator decides not to keep the tug(s) fast during the offtake, towing-off pennants suitable for quick connection to the tug(s) need to be deployed at bow and stern locations.

2.16 Lighting

Mooring, manifold and tug connection areas, including overside areas used for support vessel manoeuvring, should have adequate lighting systems to enable a safe working environment during hours of darkness. It is recommended that the illumination at the manifold and working deck areas should be in line with recommendations contained in 'ISGOTT' (Reference 1).

Consideration should be given to the provision of emergency lighting in mooring and manifold areas. This is to assist in the event of an emergency disconnection following loss of electrical power.

2.17 CCTV

The advantages of CCTV should be taken into account during the F(P)SO design stage as it enables continuous observation of manifold areas and transfer hoses/arms and effective monitoring of mooring components and relative vessel motions.

The Safe Transfer of Liquefied Gas in an Offshore Environment

Operational and Personnel Safety

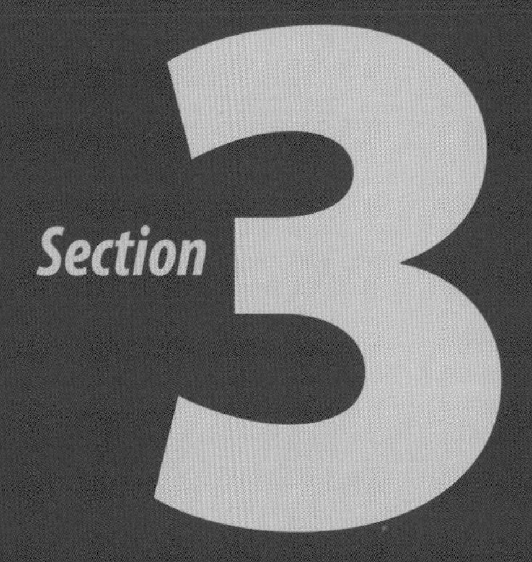

3.1 Operational Safety

3.1.1 *F(P)SO Operator's Responsibilities*

The overall management responsibility for the F(P)SO lies with the Offshore Installation Manager (OIM).

To minimise operational risk and to ensure offloading system availability, the F(P)SO operator should:

- Take full account of applicable industry guidelines
- provide trained and competent marine personnel, including a Mooring Master, Assistant Mooring Master(s) and tug and support vessel personnel to safely conduct operations
- develop procedures for the F(P)SO, including and documenting all F(P)SO marine operations procedures with a specified periodic review schedule
- develop an F(P)SO regulations manual, often referred to as a 'port information' document, to provide information to gas tankers. This should be subject to a specified periodic review schedule (see Appendix A)
- ensure that a process for gas tanker vetting and acceptance criteria/ approval by the F(P)SO is in place
- verify gas tanker fit at the F(P)SO
- provide an emergency response plan for the F(P)SO and a pollution prevention plan
- provide the gas tanker with weather forecast for duration of transfer operations
- provide appropriate material safety data sheets for the products handled
- ensure that adequate support vessels are available for conducting operations
- provide the F(P)SO and field information to the gas tanker's Master in advance of the transfer
- provide information on the F(P)SO security status to the gas tanker's Master.

3.1.2 *Scope and Content of F(P)SO Terminal Handbook*

The F(P)SO's Terminal Handbook should include national and port authority requirements, and many of the safety issues associated with the specific F(P)SO transfer operation.

An example content for a Terminal Handbook (port information booklet) is contained in Appendix A.

3.1.3 *Simultaneous Operations*

Simultaneous operations involving the F(P)SO can impede one another, presenting hazards that must be appropriately managed to minimise risk.

A matrix should be established to provide guidelines for restrictions on activities that may be conducted simultaneously with established, ongoing production operations. These should be appropriately managed through a formal management of change process. The matrix is designed to maintain the same level of safety during simultaneous activities as would be present under normal operation. An example of a SIMOPS matrix is contained in Appendix B.

3.1.4 *Communications*

3.1.4.1 *Overview*

To ensure the safe control of operations at all times, it should be the responsibility of both parties (F(P)SO and gas tanker) to establish, formally agree in writing and maintain a reliable communications system that manages the various communications associated with the F(P)SO transfer operation. This should particularly relate to communications for:

- Operations onboard the F(P)SO
- operations onboard the gas tanker
- the F(P)SO and the gas tanker

- the F(P)SO and tug(s)/support vessel(s)
- the gas tanker and tug(s)/support vessel(s).

To avoid any misunderstanding between any of the parties associated with the operation, a common language for communication, such as the Standard Marine Communication Phrases using the English language, should be used.

All communications systems, e.g. telephone, portable VHF/UHF and radio telephone systems, should comply with the appropriate safety standards and requirements. Backup systems should be provided.

The primary means of communication between the F(P)SO, tug(s)/support vessel(s) and gas tanker is UHF or VHF radio, either handheld or base stations.

On F(P)SOs, due to the complexity of the structures, VHF signals are often blocked out by the amount of additional steelwork on the topsides. UHF is, therefore, preferred.

Should communication breakdown occur, the agreed emergency signal should be sounded and all transfer operations in progress should be suspended immediately and until satisfactory communications have been re-established.

3.1.4.2 Risk Management of Critical Activities

To minimise operational risk, the F(P)SO operator should provide trained and competent personnel, including a Mooring Master and an Assistant Mooring Master(s), to safely conduct operations.

The Mooring Master should be proficient in pilotage operations and LPG cargo handling operations. The Assistant Mooring Master(s) should be proficient in LPG cargo handling operations. They should be located on site during the various critical phases of the operation, such as:

- The securing or letting go of tug(s) - an Assistant Mooring Master should be present at each location on the gas tanker that tug lines are being handled
- approach, mooring and unmooring operations - the Mooring Master should be located on the bridge of the gas tanker, one Assistant Mooring Master should be located on the forecastle and one Assistant Mooring Master should be located on the poop deck, i.e. both Assistant Mooring Masters should maintain a location together with the gas tanker's fore and aft mooring teams
- transfer hose/arm connection and disconnection operations – an Assistant Mooring Master should be located at the gas tanker's manifold
- commencement and completion of cargo transfer operations – the Mooring Master should be located in the gas tanker's cargo control room and an Assistant Mooring Master should be located at the cargo manifold
- for the duration of the cargo transfer, a Mooring Master or Assistant Mooring Master should be located in the gas tanker's cargo control room.

3.2 Personnel Safety

3.2.1 F(P)SO Manning

A description of the F(P)SO organisational structure should be included in the F(P)SO's management system. It should indicate the incumbents of key positions, and state their competence and selection criteria.

The objective of the F(P)SO manning structure is to:

- Comply with the applicable regulatory and/or national requirements
- ensure all operations are conducted safely and appropriately.

3.2.2 Training of F(P)SO Personnel

Company, local and/or national regulations will influence F(P)SO personnel training requirements.

A matrix depicting the approved and agreed training requirements for each position onboard the F(P)SO, including Mooring Masters and Assistant Mooring Masters, together with laboratory technicians, is usually maintained in the F(P)SO's management system.

In developing training requirements for F(P)SO personnel, reference should be made to the guidance and resources contained in the OCIMF publication 'Competence Assurance Guidelines for F(P)SO Personnel' (Reference 11).

The F(P)SO operator should provide adequate trained, competent and experienced personnel to support the F(P)SO cargo transfer operations and necessary personnel to board the gas tanker and to remain on board during the entire mooring, unmooring, and cargo transfer operation.

3.2.3 Competence Assurance of Key Personnel involved in Operation

3.2.3.1 F(P)SO OIM and Operational Personnel

The guidance provided in the OCIMF publication 'Competency Assurance Guidelines for F(P)SO Personnel' should be referenced when considering the competence requirements for the OIM and other operational personnel.

3.2.3.2 Mooring Master and Assistant Mooring Master

The responsibilities of the Mooring Master and Assistant Mooring Master should be prescribed in the F(P)SO operating procedures. The F(P)SO operator should provide adequately trained, competent and experienced personnel to support cargo operations, and the necessary personnel to board the gas tanker and remain on board during the entire mooring and cargo offloading operation.

Operators should ensure that the Mooring Master, notwithstanding their level of previous experience, undergoes specific F(P)SO simulator training to familiarise themselves with the range of gas tanker types expected to moor and unmoor at the F(P)SO in expected environmental conditions. In addition, there should be a continuous refresher training process during the operational life of the F(P)SO. Where an Assistant Mooring Master acts as a substitute for the Mooring Master, appropriate training should be provided including simulator training.

3.2.3.3 Personnel on Tugs/Support Vessels

The tug/support vessel should be vetted prior to being chartered for supporting the F(P)SO operations and the manning, training and competence of personnel will need to be reviewed as part of this assessment to ensure satisfactory compliance with a safety management system based upon International requirements and industry recommendations.

Consideration should be given to the benefits of tug masters receiving simulator training in support of their role in the SBS operation.

Tug and support vessel personnel should be familiar with the particular operation and this is achieved by:

- Reviewing the F(P)SO Terminal Handbook
- complying with approved and specific F(P)SO operational procedures
- holding regular toolbox meetings with all key personnel prior to all operations and at any other times deemed necessary.

3.2.4 Gas Tanker Personnel

The gas tanker will be vetted prior to being chartered and issues relating to manning, training and competence will have been reviewed as part of this assessment to ensure satisfactory compliance with a safety management system based upon International requirements and industry recommendations.

The F(P)SO Regulations contained within the Terminal Handbook are also designed to confirm the satisfactory compliance with a safety management system based upon International requirements and industry recommendations upon the gas tanker's arrival at the F(P)SO. Furthermore, a degree of familiarity with the

F(P)SO operation by key gas tanker personnel is required prior to arrival at the F(P)SO, and this can be achieved by reviewing the published Terminal Handbook.

The Safe Transfer of Liquefied Gas in an Offshore Environment

Nomination and Pre-Arrival Procedures

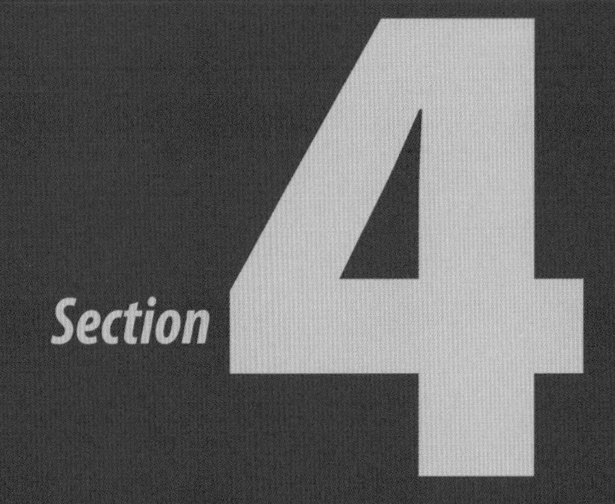

4.1 Nomination Procedures

This chapter details the processes to be followed in nominating a gas tanker for operation at an FPSO and the recommended pre-arrival communications, together with checks and tests of essential equipment.

4.1.1 Vetting

Every terminal should have a procedure in place to ensure that gas tankers accepted to call at the terminal meet minimum standards of safe operation, as established by either the terminal and/or the company's vetting department.

The process of vetting gas tankers is normally carried out by dedicated departments within shipping companies, or by third party companies contracted to undertake the vetting process. Vetting confirms the gas tanker's suitability from a safety perspective and does not necessarily confirm that it will physically fit a particular terminal, for which see Section 4.1.2.

There may be occasions where a terminal has its own processes for carrying out vetting and confirming berth fit. Regardless of the process employed, it should be ensured that a gas tanker is vetted and also assessed for facility/tanker compatibility.

4.1.2 Facility/Tanker Compatibility

Terminals should have in place a documented procedure for assessing the compatibility of nominated gas tankers. Guidance is available within the OCIMF publication 'Marine Terminal Baseline Criteria and Assessment Questionnaire' (Reference 13) and the ICS/OCIMF publication 'Ship to Ship Transfer Guide – Petroleum' (Reference 3) covering the dimensional criteria that should be assessed before terminal clearance is given. Should the terminal have location-specific criteria, a customised list should be developed detailing all relevant criteria to be assessed.

Appendix C provides an example of a gas tanker nomination questionnaire.

4.2 Pre-Arrival Communications

The 'International Safety Guide for Oil Tankers & Terminals – ISGOTT' (Reference 1) makes recommendations on communication requirements, both pre-arrival and at the ship/shore interface. The Terminal Handbook should detail the information to be exchanged with the gas tanker.

It should be ensured that the following criteria are amongst those detailed in the pre arrival exchange:

- ETA detailing the required reporting frequency
- Local Authority requirements (Customs, Immigration, Port Health). The local agent, if appointed is the usual route for such communications
- pilot boarding area
- recommended communication channels
- security status
- pre-arrival questionnaire completed
- mooring requirements, including any moorings from the terminal to the gas tanker
- manifold requirements (which side alongside, number and size of connections. State if hose or hard arm)
- tug availability and use
- SWL of ship's fitting used for securing tug lines
- weather forecast and berthing prospects.

4.2.1 Transfer Area

The Terminal Handbook should contain information relating to the transfer area that includes:

- A layout of the offshore field, including the location of all surface equipment and subsea pipelines
- the designated anchorage area, if applicable
- the pilot boarding area
- safety zones and restricted areas
- area for transfer of equipment and personnel pre-berthing, if not at the pilot boarding area
- position for making fast tugs, if applicable
- local reporting requirements
- a system for promulgation of navigational warnings to field traffic and other vessels navigating in the vicinity of the field
- security requirements approaching and within the field.

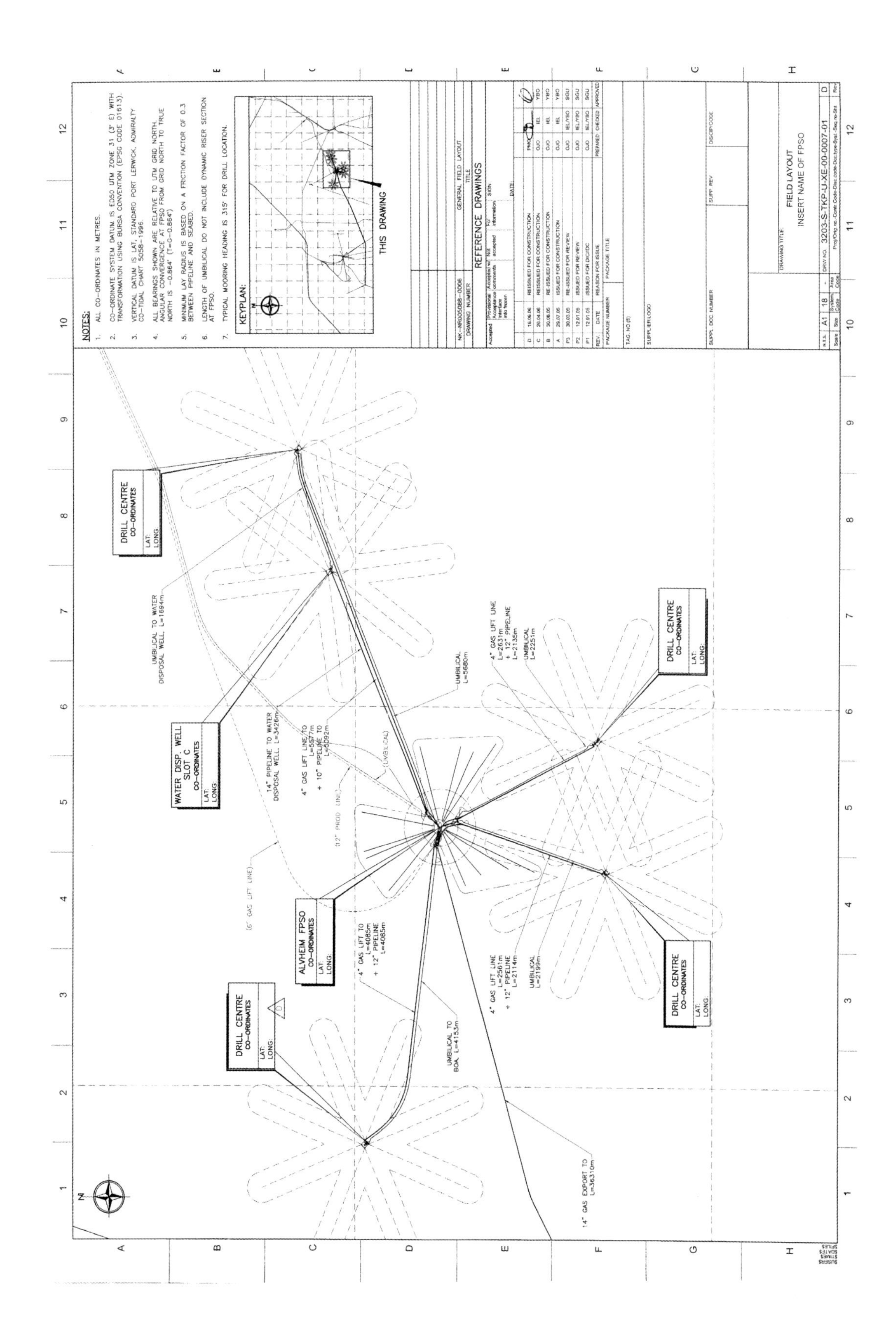

Figure 17: Example Graphic of Field Layout

4.3 Pre-Arrival Checklists and Testing of Essential Systems

The Terminal Handbook should include location-specific, pre-arrival checklists for both the terminal and the gas tanker.

The gas tanker should confirm to the terminal satisfactory completion of their pre-arrival testing of essential onboard systems.

Navigation and cargo systems to be tested should include:

- Steering gear
- main engines (including completion of trial engine manoeuvres)
- communications equipment
- control systems
- anchoring and mooring equipment
- cargo pipelines
- reliquefaction plant
- cargo hose handling equipment
- telemetry systems for ESD and emergency release, if applicable.

The terminal should carry out pre-arrival checks of essential equipment, including:

- mooring aids
- mooring equipment tested, rope messengers available
- thrusters, if applicable
- cargo transfer hoses and associated equipment
- load monitoring equipment
- cargo transfer equipment
- communication systems
- field support craft advised of gas tanker ETA and all systems confirmed operational
- weather forecasts obtained for the period of the transfer.

4.4 Tugs and Support Vessels

The gas tanker should be advised of the number and capability of tugs and support craft to be employed in the berthing operation, including their mode of operation, towing from the bow or stern of the tug and whether on a tow line, ie or pushing alongside.

If a tug is to be used in a passive or active towing mode, this should be communicated in the pre-arrival message.

Figure 18: In-Field Support Vessel and Line Handling Boat

4.5 Mooring and Fendering Plan

On the basis of the terminal design, generic mooring plans should be developed for the range of gas tankers expected to operate at the facility and copies of the plans should be included in the Terminal Regulations. The Mooring Master should develop a specific mooring plan for the nominated gas tanker, which should include:

- The running order of moorings, using a numbering or lettering system, including any moorings running from the terminal to the gas tanker
- the final mooring arrangement. This should be stored in a database on the F(P)SO for future use along with the gas tanker's basic port performance details
- details of the securing arrangements on the terminal and gas tanker, as appropriate. If grommets are to be used, details of how these are rigged, and for which moorings, should be specified
- details of the quick release arrangements provided, either onboard the F(P)SO or the gas tanker
- procedure for emergency disconnection and unmooring. This will be confirmed during the pre- transfer ship/shore conference and during completion of the ship/shore safety checklist
- confirmation of the fendering arrangement provided between the vessels and whether any fenders will be deployed from the gas tanker. The Mooring Master should ensure that the gas tanker has sufficient fairleads and bitts for the planned mooring and fendering arrangement and that these are of sufficient SWL
- details of the mooring line load monitoring system, if applicable.

Secondary fenders for the shoulder and quarter onboard the visiting gas tanker can be landed in advance en-route to the terminal or upon arrival. There should be adequate strong points onboard the visiting vessel to rig such fenders.

The mooring plan should be forwarded to the nominated gas tanker before arrival and agreement to the arrangement obtained from the gas tanker's Master.

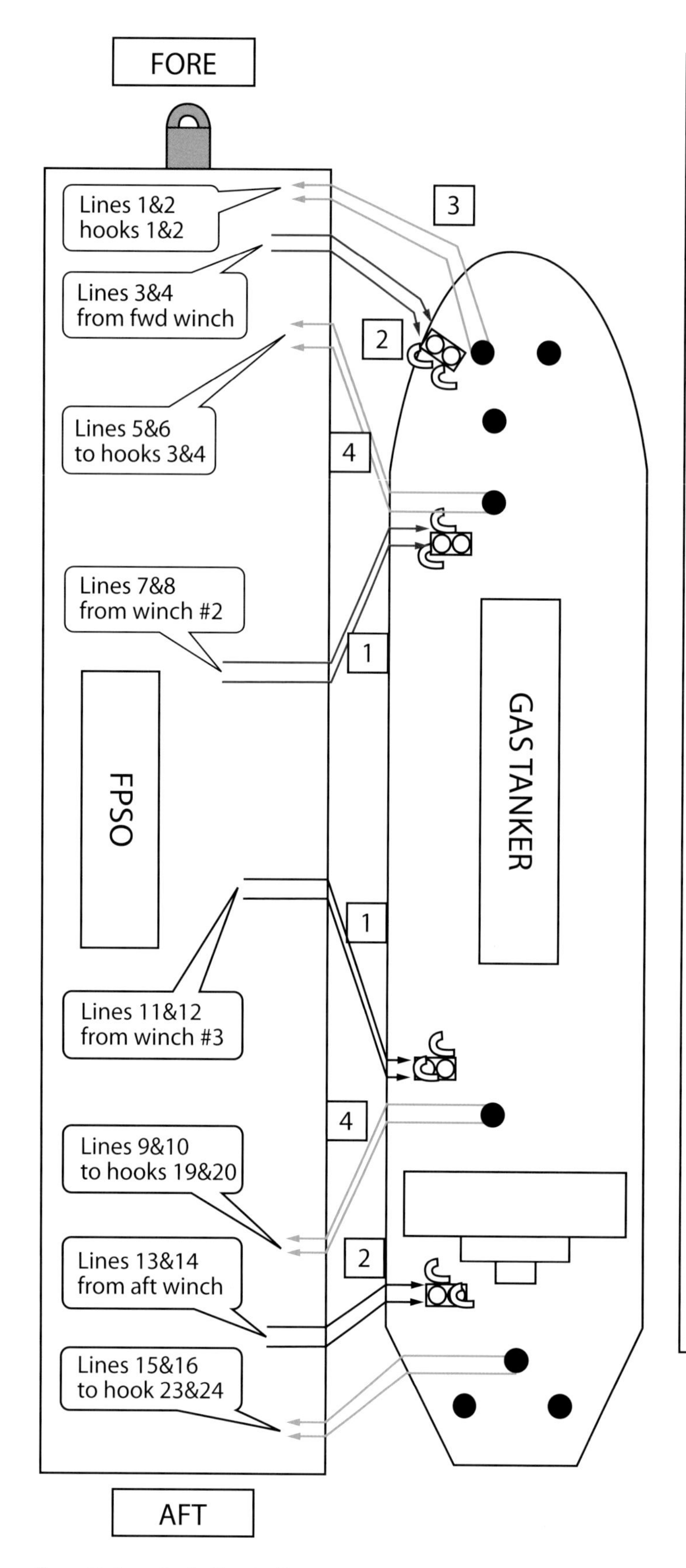

Mooring Sequence

Fore & aft sequence to be carried out simultaneously

Forward

1. Fwd springs #7 & #8 from FPSO winch #2 to be passed by rocket, both onto QR hooks.

   Split in different panama leads to avoid chafing.

   Vessel will then land on fenders.

2. Headlines #3 & #4 from FPSO fwd winch both on QR Hooks.

3. Head liines #1 & #2 from gas tanker one by one to hooks #1 & #2.

4. Lines #5 & #6 from gas tanker to hooks #3 & 4 on FPSO.

Aft

1. Aft springs #11 & #12 from FPSO winch #3 by rocket line, both onto QR hooks.

   Split in different panama leads to avoid chafing.

   Vessel will then land on fenders.

2. Stern lines #15 & #16 from FPSO aft winch to QR Hooks.

3. Stern lines #13 & #14 from gas tanker to hooks #23 & #24.

4. Spring #9 & #10 from gas tanker to hook #19 & #20.

Figure 19: Example of a Mooring Plan

The Safe Transfer of Liquefied Gas in an Offshore Environment

Berthing and Mooring

5.1 Personnel and Equipment Transfer

5.1.1 Transferring Personnel

In general, it is recommended that the transfer of personnel in an offshore environment is kept to an absolute minimum. In the case of the Mooring Master and Assistant Mooring Master(s), transfer to the gas tanker will be necessary.

All personal transfers should be conducted in accordance with current SOLAS requirements and industry guidance, such as that included in IMCA, ICS and OCIMF guidelines. Furthermore, all personnel and equipment transfers should be conducted using approved procedures.

When the gas tanker is alongside the F(P)SO, the transfer of personnel between the F(P)SO and the gas tanker may be carried out by either utilising the terminal crane and lifting equipment, provided such equipment is certified for personnel transfer, or by other equivalent means.

When the transfer of personnel is necessary prior to the gas tanker berthing alongside, a risk assessment should be conducted addressing the hazards associated with transferring personnel in an offshore environment. Such considerations should include the environmental conditions at the location and the experience and fitness of transferring personnel. Transfers during the hours of darkness should be separately risk assessed.

Where a boat transfer is being used, this should be conducted in boats that are appropriate for the task. For example, they should be small, manoeuvrable and adequately fendered with the ability to come alongside a gas tanker to transfer personnel by means of a pilot ladder, or a combination of pilot and accommodation ladders rigged from the gas tanker.

Alternatives to consider include boarding at a previous port, carrying out the transfer by boat in a more sheltered inshore location or transferring the personnel by helicopter, for which reference should be made to the guidance provided in the ICS publication 'Guide to Helicopter/Ship Operations' (Reference 12).

Basket transfers should only be carried out using cranes and equipment that is:

- Suitable for the task
- equipped with adequate safety devices to prevent free fall
- inspected and certified for personnel transfer
- operated by trained and competent personnel.

Figure 20: Examples of Personnel Transfer Equipment

5.1.2 *Transferring Equipment*

Terminal-provided equipment necessary for the SBS mooring operation is usually transferred at the same time that the terminal personnel are transferred. Equipment not required until after completion of the mooring operation should be transferred once the mooring operation is complete. Equipment necessary for the approach and the mooring operation, such as berthing aids, should be transferred to/from the gas tanker using a support vessel suitable for the task. The equipment should be transferred within its own lifting container that is appropriately inspected and certified and of a size that can be easily handled by the receiving ship.

Particular care should be taken when landing the equipment in a restricted area onboard the gas tanker.

5.2 Pre-Berthing Inspection

When arriving gas tankers are conducting operations at the F(P)SO for the first time, the terminal should arrange a physical inspection of the ship's mooring equipment. Terminal procedures may require such an inspection of every gas tanker prior to the operation.

The pre-berthing inspection should confirm the condition of all the equipment to be used for mooring, including messengers, mooring lines and mooring winches, and should also address the arrangements for securing and releasing moorings for both routine and emergency unmooring.

Any faults identified during the inspection should be rectified prior to berthing.

5.3 Pre-Berthing Information Exchange

Prior to commencing the berthing operation:

- A pre-berthing meeting aboard the F(P)SO with the F(P)SO cargo and mooring operations team, Mooring Master, Assistant Mooring Masters and tug Masters should be held to ensure that all parties are aware of their responsibilities and overall operations
- the role of the Mooring Master and the gas tanker's Master should be clearly defined in the Terminal Handbook and should be discussed during the initial meeting
- there should be discussions onboard the gas tanker between the Mooring Master and the ship's Master with regard to the conduct of the overall operations, including other simultaneous operations that may be ongoing during the period of the transfer and the berthing/unberthing operations
- the ship's Master should confirm, by means of a formal briefing, that crew members involved are familiar with the operation, particularly with regard to mooring and unmooring and including the necessary safety precautions.

During the above discussions, the role and operation of the quick release hooks and load monitoring equipment should be clearly identified and discussed with all parties to ensure proper understanding and agreement as to whose responsibility it is to control line tensions.

The fact that these discussions and briefings have taken place should be formally recorded.

Prior to commencing any final approach, the Mooring Master and the gas tanker's Master should exchange information regarding the approach and berthing operation, the ship's characteristics and operational parameters. The form referenced in the Terminal Handbook should be used to document this information exchange. Current environmental conditions and berthing abort arrangements will be an important part of this exchange.

5.4 Pre-Berthing Equipment Checks

Checks and tests of equipment onboard the gas tanker should be undertaken immediately prior to commencing the approach to the F(P)SO, and should include :

- Steering gear, including manual steering engaged in sufficient time for the helmsman to become accustomed to handling
- engines, including an astern manoeuvre
- thrusters, if fitted, confirmed operational and effective at the anticipated approach speed
- navigational equipment, such as radar and speed displays.

Terminal-provided berthing aids should be set up on the gas tanker, tested and confirmed operational, prior to commencing any final approach. Where berthing aid equipment from the F(P)SO is sited onboard, consideration should be given to siting the equipment and antenna in non-hazardous areas, such as bridge wings.

Equipment checks should also be undertaken on the F(P)SO and should include:

- Load monitoring equipment is confirmed to be operational, together with any remote load monitors
- manual and remote activation of quick release hooks
- operation of mooring winches
- operation and control of thrusters and/or tugs being used for heading control.

The sides of the gas tanker and the F(P)SO should be confirmed as being clear of any obstructions such as derrick booms, cranes and personnel ladders, prior to the gas tanker commencing its final approach.

5.5 Use of Tugs

The possible limitations of the gas tanker's bitts and leads should be taken into account, particularly when conducting operations with high horsepower tugs and/or in marginal weather conditions. Further guidance on the use of tugs may be found in the OCIMF publication 'Mooring Equipment Guidelines' (Reference 2).

Figure 21: Tug Support During Berthing

Before operations commence, communications between the Mooring Master and the tugs should be established and proven effective.

A location, clear of the terminal, should be identified for making tugs fast. The method of connection should be agreed during the pre-berthing discussions and this should include whether the connection operation is conducted with the ship stopped or with it making way.

The environmental conditions at some offshore terminals may make it unsuitable for tugs to be operated in a pushing mode, so they may have to be operated on a line. Towing lines and towing equipment should be provided by the tugs.

5.6 Berthing Criteria

The F(P)SO OIM and the gas tanker's Master should ensure that environmental conditions, including limitations due to visibility, are within the established criteria for all aspects of the operation.

Applicable weather forecasts for the area and the transfer period should be obtained and the information should be taken into account before and during operations.

If a communication breakdown occurs during an approach, the manoeuvre should be aborted. Operations should not resume until satisfactory communications have been re-established.

It is recommended that berthing operations are only conducted in daylight. Mooring operations should not be carried out at night without first carrying out an additional risk assessment of the activity, taking into account the specific conditions and problems that will be encountered at night.

5.7 Manoeuvring

During the gas tanker's approach to a weather vaning F(P)SO, the F(P)SO's heading should be oriented to provide the best berthing approach, taking into account the prevailing weather conditions and the need to prevent snatch loading on fender pennants. A careful watch should be kept on the heading of the F(P)SO to ensure that it remains steady and the F(P)SO should advise the gas tanker if she has any tendency to yaw. In the event of loss of heading control on the F(P)SO, the gas tanker's approach should be aborted until it is re-established.

The approach and manoeuvring of the gas tanker to the terminal should be as agreed during the pre-berthing information exchange.

To avoid damage to any submarine pipelines, the gas tanker's anchors should not be dropped within the field and should be confirmed fully home and secured while in the field.

Figure 22: Gas Tanker on Final Approach to an F(P)SO

5.7.1 Thruster Influence

The F(P)SO may utilise thrusters for heading control. The possible effect of thruster wash must be taken into account during the approach of the gas tanker to the terminal. Thruster wash from tunnel thrusters will be more of a consideration than wash from pod thrusters.

5.8 Mooring

Before mooring operations commence a number of rope messengers should be prepared and made available on both the gas tanker and the F(P)SO.

The order of passing lines should be in accordance with the agreed mooring plan. The first lines will be passed to the F(P)SO by the gas tanker either via a line handling boat or by use of non-pyrotechnic line throwing equipment on the F(P)SO.

When using line throwing equipment, both crews should be advised beforehand and warned again immediately before the equipment is used.

Once the first lines have been passed it is likely that subsequent lines will be passed by heaving line and messenger.

Lines should only be led through closed fairleads.

If the primary fenders are found to be incorrectly positioned, the gas tanker should be moved clear of the F(P)SO and make a new approach once the fenders have been adjusted.

The eyes of the gas tanker's lines should be placed on quick release mooring hooks on the F(P)SO. Line tensions should be monitored by the F(P)SO during mooring and throughout operations. The monitoring of load information should also be arranged to include, where used, those lines deployed from the terminal. Particular emphasis should be placed on monitoring loads immediately after mooring is complete and after any significant changes in environmental conditions or heading.

On completion of mooring, the status of tugs and thrusters should be determined in accordance with the F(P)SO's Terminal Handbook. Issues to take into consideration will be the environmental conditions, the forecasted weather and the experience that has been gained of the operation.

In view of the tug's role in the removal of the gas tanker in an emergency, consideration should be given to keeping them secured throughout operations. Where the tugs are to remain secured it is important that the length span of the tow lines, the position of the tug and the tension all be agreed.

If the tugs are released the amount of notice required for them to re-connect must be defined.

Where required, towing off pennants should be deployed in line with terminal requirements.

Any adjustment of lines during the period of operations should only be conducted with the agreement and knowledge of both parties.

Figure 23: Gas Tanker Mooring to F(P)SO

5.9 Transfer of Personnel and Equipment Alongside

Once the gas tanker is safely moored alongside the F(P)SO it may be necessary for personnel to be transferred from the F(P)SO to the tanker. The number of personnel transferred should be kept to the minimum required for operational activities.

The gas tanker will need to provide a clear area for landing the personnel and equipment. An Assistant Mooring Master should be designated as having responsibility for directing the operation as they are likely to be familiar with the type of equipment used and any safety precautions required of it.

A sufficient number of taglines of a suitable length should be used to control the lift. Under no circumstances should any taglines be secured to deck fittings while the transfer equipment is attached to a crane.

Figure 24: Transfer of F(P)SO Personnel to Gas Tanker

5.10 Orientation of F(P)SO for SBS Operations

It is important that maximum rolling limits and line tensions for safe operation are established. These limits should take into account that the rolling will be affecting both vessels, possibly in different ways, and recognise that rolling is likely to lead to the parting of mooring lines.

The terminal will normally lie orientated to the prevailing environmental conditions. However, there may be occasions where the minimum power heading is across the direction of the sea and/or swell. For SBS operations, the prime consideration is to reduce the influence of rolling between the two vessels, so the orientation of the terminal may need to be adjusted by the use of tugs or thrusters, or a combination of both, to match, for example, prevailing swell conditions. The Mooring Master or his assistant should coordinate any use of thrusters or tugs to maintain heading control and mooring integrity.

The operation of the tugs and their orientation with regard to sea and swell should not be overlooked and will need to be fully considered. This will ensure that the tugs can safely operate and remain effective throughout operations.

Providing the ship and terminal are of a similar length, pitching is not likely to be an issue as the vessels will move in a similar way, although operations should be avoided where the pitching is excessive.

Parameters should be established to ensure that the F(P)SO can maintain position and heading with the gas tanker alongside and still be able to recover heading in the worst case scenario. The method of maintaining heading control should have sufficient redundancy.

The Safe Transfer of Liquefied Gas in an Offshore Environment

Cargo Transfer

6.1 General

6.1.1 *Pre-Transfer Checklists*

Before transfer operations begin, pre-operational checks must be thoroughly discussed and carried out. Appropriate information exchange is required and the relevant parts of the Ship/Shore Safety Checklist should be completed (see Appendix D3).

A pre-transfer meeting, where the procedures within the Terminal Handbook are discussed, should be held to facilitate this exchange of information. Particular attention should be paid to:

- The setting of cargo tank relief valves and high pressure and level alarms
- remotely operated valves
- reliquefaction equipment
- gas detection systems
- alarms and controls
- the maximum transfer rate
- agreeing on who will connect the cargo transfer system
- whether or not scuppers will be plugged
- method for draining and clearing cargo hoses at the end of the discharge.

These checks should be carried out while taking into account any known restrictions of gas tanker or F(P)SO systems. Any special precautions for specific cargoes should be made known to personnel onboard the gas tanker.

The gas tanker should be advised of the F(P)SO's policy on suspending operations during electrical storms. Marine terminals handling bulk hydrocarbon cargoes will usually stop all transfer operations when an electrical storm is approaching or is in the vicinity. However, it is acknowledged that conditioning is such is that under normal operating conditions there should be no hydrocarbon emissions. Any decision for the F(P)SO to continue transfer operations during an electrical storm should be based on a formal risk assessment and will depend upon the design and equipment fitted to the both the F(P)SO and the gas tanker.

6.2 Cargo and Ballast Plans

6.2.1 *Cargo Plans*

A cargo transfer plan should be agreed with the Responsible Officer during the pre-transfer meeting. The transfer rate and the total amount of cargo to be transferred should be defined in the plan, together with the cargo tank pressures for the gas tanker. The plan should also address issues that may include the following:

- Cargo density, temp and pressure
- loading rates and any limiting conditions for cargo tanks (sloshing) and including the use of booster pumps (LPG)
- dealing with leaks and spills – vapour and liquid
- emergency cargo transfer shutdown protocol, including the management of liquid lock following an ESD
- the generation of vapours caused by vessel movement
- if vapour return lines are to be supplied
- custody transfer procedures

- tanks to be used for cargo parcel. (Note: production loading on the F(P)SO should be segregated from those tanks being discharged. Therefore, there is generally a need to ensure adequate capacity for anticipated produced volumes, which should be a consideration at the cargo planning stage).

6.2.2 *Ballast Plans*

Ballasting and de-ballasting plans should be drawn up by both the F(P)SO and the gas tanker. The ballasting plan and the anticipated freeboard changes should be discussed by the Mooring Master and the Master of the gas tanker. For SBS mooring systems, to avoid any undue stress on the mooring lines, the changes in freeboard between the two vessels should be kept to a minimum, subject to safe operations. Generally, large differences should be avoided as this increases the vertical orientation of the mooring lines, adversely affecting their efficiency.

Both the F(P)SO and the gas tanker should be upright with list and trim maintained within operational limits throughout the mooring, cargo transfer and unmooring operations.

6.3 Pre-Transfer System Integrity Checks

The following conditions are required before the start of any cargo transfer operation:

- All safety systems, plant and equipment routine checks have been completed
- ESD system has been proven operational
- all valves, controls and instrumentation for the cargo and cargo support systems are to be checked for correct operation and any necessary repairs or adjustments carried out
- valve status is as listed in operating procedures
- where fitted, variable setting pressure relief valves, high tank pressure alarms and gas detection sample valves should be correctly set
- the pre-start checks and calibrations for the portable gas analysers should have been performed
- tools needed for the operation must be available.

6.4 Hose/Arm Handling and Connection

6.4.1 *General*

A visual inspection of each hose assembly should be carried out, before it is connected to the manifold of the gas tanker, to determine if any damage has been caused while taking them onboard. If damage to a hose or flange is found that impacts on their integrity, the hose should be withdrawn from use.

During operations, the proper handling of hoses is important and hoses of all types should be correctly supported when connecting, during cargo transfer and during disconnection. Care should also be exercised when rigging or moving hoses to ensure that they are not damaged or laid against sharp edges that could weaken the hose. A recognised good practice is to protect manifold flanges with plywood blanks during hose handling.

Transfer hoses should be checked for leakage. This is commonly achieved using soapy water to check the connections. The crew of the gas tanker should do the same on their manifold.

Wrapping the connecting flanges of the hose and manifold in plastic before the start of cargo transfer will prevent ice formation directly on the flanges, bolts and nuts, thereby facilitating disconnection. It will also eliminate the need to use water to melt the ice from the connections.

Figure 25: Handling of Transfer Hoses

6.4.2 Drying and Inerting the Transfer System

If transfer lines, hoses or arms have been open to the atmosphere, it will be necessary to dry and inert the system prior to connection.

6.4.2.1 Transfer System Drying

All water must be removed from the transfer system to avoid icing and hydrate formation. Whatever method is adopted for drying, care must be taken to achieve the correct dew point temperature.

6.4.2.2 Transfer System Inerting

If the transfer system has been open to the atmosphere it will require inerting to ensure that there is a non-flammable condition during the subsequent gassing-up of the transfer system with cargo. For this purpose, oxygen concentrations must be reduced from 21% to a maximum of 5% by volume, although lower values are often achievable.

6.4.3 Bolted Flanges

Bolted flange connections are often used for offshore LPG transfers. The control of the movement of the hose/hard arm is critical to ensure flanges are kept properly aligned during connection. The condition of flange surfaces should be checked prior to connection.

6.4.4 *QC/DC Couplings*

All types of QC/DC couplings need to be carefully checked for their mechanical integrity and function prior to connection. Locking faces and 'O' rings should be checked for damage.

QC/DC couplings may require further tightening once the system is cooled down.

6.4.5 *Emergency Release Couplings (ERCs)*

ERCs should be checked to ensure that they are in an operable condition and ready for use. For example, any arrangements for securing the ERC when not in operation should be removed once the hose/hard arm is connected.

ERC support systems (e.g, pneumatic, hydraulic, nitrogen) should be fully available throughout transfer operations.

6.4.6 *Loading Arms*

If hard arms are employed, the manufacturer's procedures for their operation should be adhered to.

As hard arms may still be in gear under hydraulic power at the time of connection and before being dis-engaged to 'free-wheel' with the ship, operators should be aware of the potential for high loads to be imposed on the gas carrier's manifold.

6.4.7 *Leak Testing the Transfer System*

Once hoses or arms are connected to the gas tanker, and prior to commencing cool down, the transfer system should be checked for leaks. This is normally undertaken using an inert gas under pressure, nitrogen for example. Particular attention should be given to the integrity of connections at both manifolds.

6.5 Sampling and Gauging

6.5.1 *General*

Fiscal measurement of the quantity and quality of cargo transferred will normally be governed by standards maintained by the field operator/cargo owner, international and local regulatory bodies, and accepted codes of practice, such as those published by ASTM. F(P)SO operators will have strict procedures that comply with these standards.

6.5.2 *Product Measurement*

The quantity of liquefied gas loaded onto a gas carrier can be determined by conventional tank gauging in the usual manner. However, the cargo/vapour measurement system onboard the F(P)SO can consist of a number of subsystems or components for each cargo tank, such as:

- Hydrostatic tank gauging
- radar ullage measurement
- magnetic float gauging
- automatic tank thermometer.

Alternatively, the volumes transferred can be monitored by flow meter. It should be noted, however, that where vapours are returned to the tank being discharged, their quantity should be taken into account in the cargo calculations unless otherwise specified.

On the F(P)SO, tanks scheduled for discharge should be separated from any process activities or any refrigerated or reliquefied returns.

6.5.3 *Sampling*

F(P)SOs may operate an automatic line sampler ('grab' sampler) during a cargo transfer. This equipment can be adjusted for the frequency and size of 'grabs' during a transfer to correspond to the total cargo to be transferred and the expected transfer time. Alternatively, representative liquid tank samples will be taken manually from the F(P)SO tanks scheduled for discharge.

6.6 Cargo Transfer

Cargo transfer activities follow normal practice for operations at any marine terminal with added considerations, not limited to the following:

- Vapour generation during initial stages of loading due to gas tanker movement, particularly rolling
- maximum transfer rate based on reliquefaction capacity of gas tanker
- actions to be taken in the event of reliquefaction plant failure
- the implications to cargo transfer in the event of having to change-over tanks on the F(P)SO
- the cargo plan should define the party responsible for stopping cargo on receipt of nominated quantities and the manner of ramping down and stopping transfer
- checks on quantity transferred before line clearance and hose disconnection.

6.7 Control of Vapours

Due to the proximity of the F(P)SO's processing plant, it is important that tank pressures on the gas tanker are controlled to ensure hydrocarbon venting does not take place. The control of cargo vapours during loading can be carried out using:

- A vapour return line to the F(P)SO coupled to a gas compressor
- the ship's reliquefaction plant for liquid return to the ship's tanks
- a combination of the above.

For fully refrigerated gas tankers, a vapour return line may be connected to the ship's vapour manifold, but this is most often put in place for safety reasons to provide pressure relief. The normal loading practice on such ships is to load through the liquid header, while drawing off excess vapour via the vapour header, operate the reliquefaction plant and return liquid to the ship's tank via the condensate return line. If necessary, loading rates should be adjusted to enable the reliquefaction process to maintain the required tank pressures. Should tank pressures become excessive, the ship's vapour manifold valve can be opened to relieve the situation by returning the vapours to the F(P)SO.

If the gas tanker is taking a cool-down parcel from the F(P)SO, care must be taken, if venting the inert gas, to prevent an inert gas/ hydrocarbon gas mixture being released. The tanker should only take the quantity of cool-down liquid required and then loading operations should stop. The tanker should then unberth from the F(P)SO and move to a safe area clear of the production field and complete the cool-down process where it is safe for hydrocarbon venting to take place. Once cooled down with all inert gases vented from the system, the tanker can re-berth alongside the F(P)SO and complete loading in the normal manner.

6.8 Hose/Arm Clearing

6.8.1 General

On completion of cargo transfer, the transfer system should be drained of liquid, under gravity, to both the gas tanker and the F(P)SO. The transfer system must then be made liquid-free by using hot gas or nitrogen, as detailed below.

The procedures for draining, hot gassing or nitrogen purging the cargo transfer system should be properly evaluated and discussed during the pre-transfer briefing. Factors affecting the choice of the party providing the hose/arm clearing medium may include the following:

- F(P)SO procedures
- expected pressure in tanks on completion of cargo transfer
- grade of cargo handled
- production and/or internal cargo transfer operations onboard the F(P)SO at the time of hose blowing
- type of gas tanker
- freeboard differences
- previous cargoes carried by the gas tanker and the composition of her cargo tank vapour phase.

It is recommended that the F(P)SO is the preferred party for providing hot gas or nitrogen purging as it normally has more flexibility and control over most of the above factors.

The tank nominated for receiving the drained or displaced cargo from the transfer hose/arm and associated pipelines should be confirmed as having sufficient space to receive the anticipated volumes.

6.8.2 Hose Draining

Unless the system is designed to do so, it is not generally recommended that the connected cargo hose is lifted with a crane or derrick in order to drain, by gravity, any liquid remaining in the hose. Large freeboard differences and the relative movement of hulls in adverse weather may cause excessive strain or even damage to the hose, connecting flanges or lifting equipment.

However, in the case of a transfer of butane in a low temperature area, physically lifting the hose may need to be considered as the butane may not vaporise. A careful assessment of weather conditions and relative vessel movements should be undertaken prior to the bight of the connected hose being lifted.

Draining of the hose back to the F(P)SO by opening a valve on the F(P)SO cargo line leading to storage tanks will be facilitated by the high cargo pressures of a fully pressurised type gas tanker.

6.8.3 Hose/Arm Clearing with Hot Gas

It is preferable to supply the hot gas from the reliquefaction units of the F(P)SO. If the F(P)SO is not capable of supplying hot gas, the gas tanker can be used to send hot gas back along the transfer system to the F(P)SO.

Before commencing the line clearing operation, both the F(P)SO and the gas tanker should set their cargo systems. The party providing the hot gas must follow procedures established for the operation within its relevant procedural documentation. It should be ensured that the hot gas is of the same grade as the transferred cargo.

The receiving side should close their manifold valve. Clearing of the hose/arm is achieved by building up pressure in the hose/arm with the hot gas (typically up to 7 Barg) and then releasing it by opening the manifold valve. This operation should be repeated until the hose/arm is liquid free, maintaining close communications between both parties throughout.

The pressure within the tanks of a fully pressurised gas tanker may require a higher pressure of hot gas to be delivered to clear the hose/arm and, in some cases, the tanker may be the preferable party from which to blow the contents back to the F(P)SO.

6.8.4 Hose/Arm Clearing with Nitrogen

Some F(P)SOs and gas tankers have facilities for the production and use of nitrogen.

The advantage of using nitrogen for clearing and purging hoses/arms includes the possibility to achieve higher blowing pressure and a consequently shorter time for the operation. In addition, where hoses are to be disconnected from the F(P)SO manifold for storage in a rack, they will not require further inerting after the nitrogen purge.

The main disadvantage associated with using nitrogen is the introduction of non-condensable gases into the cargo tanks, resulting in reduced efficiency of the refrigerating plant. For this reason, semi-pressurised and fully refrigerated gas tankers are generally reluctant to accept nitrogen as the hose clearing medium.

If the safety and operational philosophy requires the use of nitrogen for hose clearing, the above factors should be taken into account at the F(P)SO's design stage. In particular, consideration should be given to the provision of a connection for a small diameter hose between the F(P)SO's nitrogen delivery outlet and the gas tanker's manifold drain valve, as described in Section 2.11.1.3.

6.9 Hose/Arm Disconnection and Recovery

Disconnection of hoses/arms onboard the gas tanker can be carried out by the gas tanker's crew or by F(P)SO personnel, as prescribed by the F(P)SO's operating procedures.

The disconnection operation should not be commenced until pressure in the system is released and the manifold valves on both the F(P)SO and gas tanker are closed, together with any hose end valves. Proper communication must be maintained to ensure that preparatory activities are completed on both sides.

Personnel taking part in the operation must follow safety procedures as per the product's MSDS, and appropriate PPE should be used to protect against possible cargo splashing, cold burns and contaminants.

Flanges must be fully blanked once hoses or arms are disconnected.

The hose will normally be lifted from the deck of the gas tanker using the F(P)SO's crane or derrick. The hose manufacturer's recommendations for handling, particularly in respect of minimum bending radius, should be followed. If lifting bridles are used they must be evenly placed and properly secured.

Precautions should be observed during the lifting operation to avoid injury to personnel or damage to the hose or manifold fittings and equipment on the gas tanker due to the swinging of the hose caused by adverse weather and hull movement. If deemed necessary, additional ropes should be attached to the hose to control its movement.

Hoses should be stowed and secured in accordance with the F(P)SO's procedures and the manufacturer's guidance.

Figure 26: Hose Being Secured in Storage Rack

The Safe Transfer of Liquefied Gas in an Offshore Environment

Unmooring and Departure

7.1 Early Departure Procedure

The early departure procedure allows the gas tanker to commence the voyage to the next port as soon as possible after completion of the cargo transfer and after the F(P)SO personnel have disembarked, yet before the bill(s) of lading and associated cargo documentation have been checked and signed by the gas tanker's Master. An authorised representative will sign the bill(s) of lading and associated cargo documentation on behalf of the gas tanker's Master, following the Master's written approval of the details presented.

The early departure procedure is generally considered to be of a commercial nature. Operationally however, the early departure procedure is generally viewed as advantageous as it minimises the time the gas tanker remains moored alongside the F(P)SO and/or within the general field location. The minimisation of this time impacts on the risk analysis and reduces the exposure from potential incidents. As a consequence, F(P)SO operators should carefully evaluate the commercial and operational implications associated with the adoption of early departure procedures.

7.2 Unmooring and Departure

On completion of cargo transfer and the disconnection and recovery of cargo hose(s)/arm(s), the personnel responsible for the unmooring operation should ensure that:

- All F(P)SO personnel and equipment not required for the unmooring operation are transferred to the F(P)SO
- all personnel and equipment required for unmooring are correctly located and ready for operations
- the unmooring procedure and sequence is clearly understood and agreed to.

In determining the above, reference should be made to the pre-berthing checklist (Appendix D1) and particular consideration should be given to the following:

- The prevailing weather – wind speed and direction, current speed and direction, and swell(s) height and direction
- the ability of the F(P)SO to maintain the optimal heading during the unmooring operation
- the sides of the gas tanker and F(P)SO are cleared of obstructions such as derrick booms, cranes and personnel ladders, prior to the unmooring operation
- tug(s) are secured to the gas tanker and are ready for the unmooring operation
- the F(P)SO OIM and the gas tanker's Master are aware of any simultaneous operations that may impact on the unmooring and departure operation
- other vessels/drilling rigs in the field location are aware of the unmooring operation
- fenders, including their towing and securing lines, should be checked to be in good order. If not, the unmooring plan should be adjusted accordingly
- communications are confirmed operational
- appropriate personnel are correctly located on the F(P)SO and the gas tanker
- the method and sequence of release of moorings should be in accordance with the unmooring plan. When slacking down moorings for release, it should be noted that some tension is required on the line to trip the QRHs after their release.

The unmooring sequence should allow sufficient time to ensure that all mooring lines are fully recovered prior to progressing to the next stage of the sequence. This will ensure that, should any mooring line become entangled with the fenders, the hazards are adequately managed. The sequence should also include consideration of when propulsion systems may be required.

When all mooring lines are clear of obstructions and are clear of the water surface, the gas tanker can be bodily moved away from the F(P)SO while maintaining a parallel heading. When the gas tanker is well clear of the F(P)SO it should be manoeuvred to a position that is downwind and down-current of all field facilities, as agreed during the pre-unmooring information exchange. At no stage in the operation should the gas tanker pass upwind or up-current of the F(P)SO.

Once the gas tanker is at a safe location, tug(s) should be disconnected and remaining equipment and F(P)SO personnel transferred.

The Safe Transfer of Liquefied Gas in an Offshore Environment

Gas-Up and Cool-Down

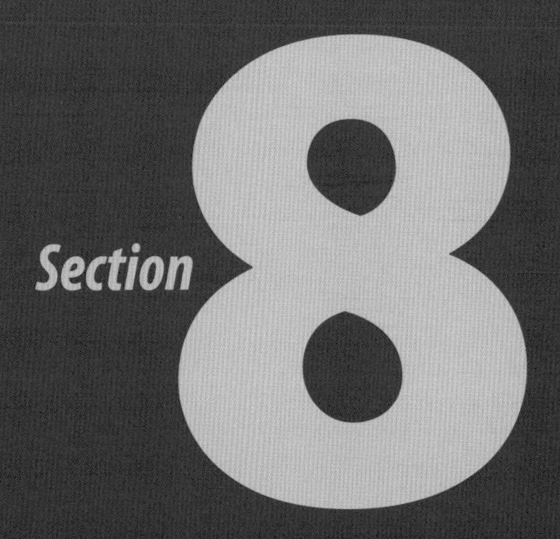

8.1 General

Under normal conditions the gas tanker arrives ready to load with her tanks under vapours of the nominated cargo grade and cooled down. Likewise, the F(P)SO is normally in a routine operational state with cargo in the tanks ready for offloading.

However, there are cases when a gas tanker may be nominated to load at an F(P)SO with her cargo tanks under inert gas. This can be when the tanker arrives from a shipyard on her maiden voyage or after a scheduled or unscheduled repair, after any emergency repair undertaken during the voyage and requiring verification of independent surveyor, or after a major grade change, for example, from ammonia or any other chemical gas to LPG. Before the nomination of an inerted tanker is accepted, the F(P)SO must confirm that the gas-up and cool-down operations can be conducted safely.

Apart from the cases where the F(P)SO Regulations specifically prohibit the acceptance of an inerted tanker, there should be clear procedures to address the safe and efficient conduct of operations.

8.2 Gas tanker

8.2.1 Gas Tanker with Tanks under Air

A gas tanker arriving with tanks aerated should not be allowed alongside the F(P)SO. All necessary cleaning, inspections and inerting must be carried out at anchor or while adrift, at a safe distance remote from the field.

Prior to the gas tanker being accepted for berthing alongside the F(P)SO, the cargo tank atmospheres should be verified by F(P)SO personnel and confirmed to be in an inert condition.

8.2.2 Gas Tanker with Tanks under Inert Gas

The most efficient way of conducting the gassing-up and cool-down operation must be identified, with the aim of minimising time alongside.

Procedures should be aimed at providing sufficient liquid to the gas tanker to enable gassing-up and cool-down of all cargo tanks. During the loading of the cool-down parcel, the gas tanker should vent the displaced inert gas through a vent riser and at no time should the release of hydrocarbon vapours be permitted. If possible, the heading of the F(P)SO should be orientated to keep any emissions from the gas tanker's vents clear of process areas. Venting through a cargo hose hanging over the side of the gas tanker should be prohibited while the tanker is alongside the F(P)SO.

Procedures for loading the cool-down parcel may include the possibility of using the F(P)SO vent-off mast or cold/hot flare for handling efflux from the gas tanker via a vapour hose connected to the F(P)SO's vapour system.

Once the cool-down parcel has been loaded the gas tanker should unmoor. It should continue gassing-up operations at a safe location clear of the F(P)SO and other field installations.

8.2.3 Loading Propane Cargo into Tanks under Butane Vapours

The main issue associated with accepting a fully refrigerated gas tanker for a propane cargo at an F(P)SO, when the previous cargo was butane, is that cool-down operations will be extended, resulting in the gas tanker being alongside the F(P)SO for a longer period. Although the gas tanker will generally be presented ready to load, tanks may be relatively warm due to the boiling temperature difference between propane and the previous butane grade (minus 42°C against minus 0.5°C).

The loading operation must be started at a very slow rate, normally to one tank only, and pressure in the tanks must be closely controlled to prevent pressure relief valves from lifting.

8.3 F(P)SO

8.3.1 Preparation and Arrival on Location

Newly built F(P)SOs, or those following a major conversion, will need to test and commission all systems before receiving acceptance from parties that may include owners, clients, Class, Flag State and/or Coastal State interests.

Providing relevant systems are ready for testing and commissioning, the operator may decide to undertake gas trials before arrival on location. In this case, an amount of LPG sufficient for gassing-up the tanks, as well as for running the gas plant and cargo pumps, can be supplied via tank barge or small vessel. On completion of the gas trials the vessel will likely undertake gas-free operations, recovering all possible liquid for return to the tank barge or supply vessel, with unrecoverable product vented into atmosphere. The advantage of this approach is that it provides the possibility to correct or repair any equipment or system while still within reach of a shipyard. Disadvantages are primarily associated with the additional costs and time for this option.

The F(P)SO may be gas free on arrival on location. Once the F(P)SO is moored and connected-up there are two main options for gassing-up cargo tanks, pipelines, process plant and gas plant. The operation can be carried out either by chartering a gas tanker and mooring it alongside or by using natural gas liquids (NGL) or LPG from a supply source such as a rig, platform or barge fitted with a fractionator.

8.3.2 Use of Gas Tanker

All process and cargo systems on the F(P)SO must be properly inerted prior to the commencement of gassing-up and cool-down.

This option allows the testing and validation of other systems on the F(P)SO, such as those associated with mooring, fendering, hose handling, positioning and orientation control. It will also provide a practical check of the Terminal's procedures and the efficiency of communications.

The gas tanker should preferably be of the most common size that the F(P)SO is designed to handle.

This gassing-up and cool-down operation requires thorough discussion and planning between both parties and ample time should be allowed for this. The aim should be to load a sufficient quantity and quality of vapour and liquid to enable the gas-up and cool-down of the F(P)SO tanks, following which the gas tanker will be unmoored.

The operation will be considerably longer than a normal operation and additional time should be allowed for the commissioning and testing of systems.

Efflux from the F(P)SO's tanks and systems will be routed to a flare, if fitted. The flare can be ignited once a sufficient concentration of hydrocarbons is reached. Should no flare be fitted, the efflux should be routed to a suitable vent riser (cold flare).

Thrusters, if fitted, can be used to maintain a favourable heading to facilitate safe dispersion of efflux from the flare or vent riser.

8.3.3 Use of Supply Source for Gassing-up FPSO

This option eliminates risks associated with having a gas tanker alongside, as well as the associated cost. However, it does not allow the testing and validation of systems required for receiving a tanker alongside.

The use of product from a supply source may impact on the specification of the first cargo to be offloaded from the F(P)SO.

The gassing-up and commissioning of the F(P)SO's cargo storage systems needs to be detailed within the overall platform start-up planning and procedures. Any issues associated with commissioning conflicts between process and storage functions should be identified at the planning stage.

The Safe Transfer of Liquefied Gas in an Offshore Environment

Offshore Transfer of LPG via Tandem Mooring or SPM

9.1 Primary References and Scope

OCIMF's publication 'Tandem Mooring and Offloading Guidelines for Conventional Tankers at F(P)SO Facilities' (Reference 4) should be the primary publication referenced for all F(P)SO tandem offloading operations.

For SPM operations a number of OCIMF publications provide guidance, including 'Single Point Mooring Maintenance and Operations Guide' (Reference 6) and 'Guidelines for the Purchasing and Testing of SPM Hawsers' (Reference 8).

The publications contain current knowledge that is pertinent to transfers involving oil, not LPG. This chapter highlights the differences associated with the safe transfer of liquefied gases in the offshore environment that are not discussed in existing OCIMF guidance or in this publication's previous chapters.

Figure 27: Offshore Transfer of LPG via Tandem Mooring

9.2 Safety in Design

9.2.1 Mooring Equipment Layout, Configuration and Compatibility

9.2.1.1 Gas Tanker Equipment

For tandem or SPM operations, incompatibility between mooring equipment and shipboard fittings can prejudice both mooring system security and personnel safety. Equipment incompatibility may also serve to prolong mooring operations. These difficulties can be overcome by ensuring that gas tankers are provided with mooring equipment in accordance with the OCIMF publication 'Recommendations for Equipment Employed in the Mooring of Ships at Single Point Moorings' (Reference 7).

Although it is recognised that gas tankers are not normally provided with equipment to enable them to moor safely in tandem at F(P)SOs or SPMs, it is recommended that gas tankers should be provided with one or two bow chain stoppers and associated bow fairleads and winches, designed and arranged in accordance with OCIMF's published guidance.

9.2.1.2 *F(P)SO Equipment*

Reference should be made to guidance provided in 'Tandem Mooring and Offloading Guidelines for Conventional Tankers at F(P)SO Facilities' (Reference 4), and relevant sections of this publication.

9.2.1.3 *SPM Equipment*

Reference should be made to the guidance provided in 'Single Point Mooring Maintenance and Operations Guide' (Reference 6), and 'Guidelines for the Purchasing and Testing of SPM Hawsers' (Reference 8).

9.2.2 *Cargo Transfer Equipment*

9.2.2.1 *Gas Tanker Equipment*

Equipment provided by the owners of gas tankers should meet the technical recommendations and guidance outlined in the latest edition of the following publications:

- 'Recommendations for Manifolds of Refrigerated Liquefied Gas Carriers for Cargoes 0°C to minus 104°C 2nd Ed' (Reference 10), with regard to the general manifold configuration
- 'Recommendations for Oil Tanker Manifolds and Associated Equipment' (Reference 9), with regard to the fittings required at the manifold to safely handle the transfer hoses offshore.

Gas tankers scheduled to undertake tandem or SPM operations should be provided with hose handling equipment suitable for handling the transfer hose, taking into account the following guidance:

- Lifting Equipment – gas tankers should be provided with lifting equipment on either the starboard or port side, with a minimum safe working load of 5 tonnes when plumbing a point one metre outboard from the ship's side over the length of the cargo manifold used to transfer cargo.

  Typically, a 300 mm hose tapered to 200 mm is used to reduce the weight of the hose end to stay within the 5 tonnes SWL limitation. Terminal operators should conduct a detailed engineering evaluation to determine the maximum dynamic load on the gas tanker's crane. (Refer to guidance in 'ISGOTT' – Reference 1)

  The gas tanker's lifting equipment should also be able to provide a clear lift above the deck of at least 8 metres to the manifold connection
- height of cargo manifold - the centre of the presentation flange should be located at least 700 mm above the horizontal projection of the top of the hose support at the ship's side.

  The height of the centre of the presentation flange above the deck should not exceed 2 metres

  Distance from ship's side - the distance of the presentation flange inboard from the ship's side should be approximately 3.5 metres (+/- 0.5 metres)
- hose support at ship's side - a way to adequately support a 200 mm or 300 mm hose in way of the ship's side abreast of the cargo manifold should be provided. A horizontal curved plate or pipe section should be fitted at the ship's side at least 700 mm below the centre of the presentation flange. The hose support should have adequate design strength to be capable of withstanding a load of 10 tonnes
- fairlead - one closed fairlead for each hose, with a clear opening of approximately 400 x 250 mm, should be provided at the ship's side in line with the manifold(s) for hose chains and hose hoisting having a minimum safe working load of 25 tonnes
- cruciform bollard - one cruciform bollard 600 mm in height, or other suitable bollard, for each hose should be conveniently placed near to the deck area between the manifold and the ship's side, allowing maximum clear area for safe access. The minimum SWL of the bollard should be 25 tonnes
- deck rings – four deck rings of 15 tonnes SWL should be fitted on the deck inboard of each fairlead, located approximately 2 metres and 3 metres forward and aft of the centre line of the fairlead.

9.2.2.2 *F(P)SO/SPM equipment*

Reference should be made to either 'Tandem Mooring and Offloading Guidelines for Conventional Tankers at F(P)SO Facilities' (Reference 4), or the 'Single Point Mooring Maintenance and Operations Guide' (Reference 6), for information on equipment to be provided by F(P)SO/SPM facilities.

For hoses used in offshore marine LPG applications, it is recommended that reference is made to the relevant guidance in the OCIMF publication 'Guide to Manufacturing and Purchasing Hoses for Offshore Moorings' (Reference 5), particularly with regard to issues associated with the hose's service in the marine environment.

9.2.2.3 *Emergency Release Coupling*

Consideration should be given to the inclusion of an emergency release coupling (ERC) in the transfer system (refer to Section 2.14).

9.2.2.4 *Marine Breakaway Coupling*

Consideration should be given to the inclusion of a marine breakaway coupling (MBC) in the hose string when conducting operations via SPM or tandem mooring.

9.2.3 Equipment Evaluation

The aim of these guidelines is to promote the standardisation of gas tanker and terminal operational equipment.

During the terminal design phase, F(P)SO and/or SPM operators should perform a site-specific evaluation of hose weights and equipment to be used during gas tanker cargo transfer operations. This evaluation should be based on environmental conditions at the specific site.

The objective of the evaluation should be to verify that OCIMF's equipment recommendations are suitable for safe operations and will not significantly restrict the operational capability of the terminal. Should the evaluation indicate that the OCIMF recommended equipment may restrict a terminal's operational capability or create significant risk, terminal operators should provide specific design recommendations for gas tanker equipment when operating at that terminal, which may create a requirement for dedicated tonnage.

9.3 Operational Issues

Generally, the information on transfer activities associated with SBS transfers of LPG will be applicable to transfers undertaken via a SPM or tandem mooring. Where differences exist they will be detailed in the referenced publications. The following are among the issues to be considered:

- For the tandem or SPM operation, all equipment and personnel required for mooring and hose connection will need to be transferred to the gas tanker prior to mooring
- gassing-up, cool-down and line clearing operations will take longer than an SBS operation
- a vapour return facility is unlikely to be provided at an SPM or tandem mooring.

The Safe Transfer of Liquefied Gas in an Offshore Environment

Emergency Preparedness

10.1 Overview

10.1.1 General

Emergencies can occur at any time and in any situation. While it is impossible to account for every eventuality, effective action is only possible with pre-planning and the development and exercising of emergency procedures.

In any operation where two entities are combined to undertake a common process, there are likely to be conflicting priorities and ideals. While both the F(P)SO and the gas tanker should have their own individual, well established emergency plans, it should not be expected that each is fully aware of the other's. It is, therefore, essential that mutually agreed procedures are available that define the action to be taken by either party in the event of an emergency.

The F(P)SO Terminal Handbook should provide information to the gas tanker on actions that will be taken by the F(P)SO and those that are required of the gas tanker. The procedures should cover all types of emergency that can be reasonably envisaged in the context of the operation, for example:

- Power failure on the F(P)SO, gas tanker or tug
- collision/heavy contact
- grounding in approaches to or within field
- loss of containment
- uncontrolled venting
- fire and explosion
- hose/arm failure
- mooring failure
- fender breakaway or burst
- medical evacuation
- helideck incident
- oil pollution
- security incident.

All personnel involved in the operation must be familiar with the emergency procedures, should be adequately trained and should clearly understand the action they are required to take when responding to an emergency. This should include the sounding of alarms, the setting up of a control centre and the organisation of personnel to deal with the emergency.

Information on the hazards associated with the products being handled should be readily available in case of an emergency. Material Safety Data Sheets (MSDS) should be provided that contain advice to personnel on handling or working with the product(s).

The availability of sufficient manpower is necessary to initiate successfully, and then sustain, any response plan. At the planning stage a study should be undertaken to determine the total manpower requirements over the whole period of any potential emergency, taking into account the hour at which the emergency may occur. An F(P)SO may be isolated and remote with little or no immediate support in the way of manpower or equipment other than its own staff. As part of a Safety Case, the needs for manpower and resources should be identified, with consideration given to the use and capability of any field service vessels and/or neighbouring field operations that may be able to assist in an emergency through mutual agreement.

Tugs and other field service vessels should be equipped with fire fighting capability and the crews should be trained and knowledgeable in relevant fire fighting techniques.

The most important and critical elements of every emergency plan are the organisation and resources necessary to support it. When drawing up the plan, all parties who are likely to be involved should be consulted, with consideration given to including potential gas tanker operators/managers.

It will be necessary to:

- Analyse probable emergency scenarios and identify potential problems
- agree on the best practical approach to respond to the scenarios and to resolve identified problems
- agree on an organisation with the necessary resources to execute the plan efficiently.

The plan should be reviewed and updated on a regular basis to ensure that it reflects any changes within the F(P)SO, current best practice and any key lessons from emergency exercises and/or previous emergencies.

10.1.2 Emergency Plan Components and Procedures

10.1.2.1 Plan Preparation

The F(P)SO's emergency plan should harmonise with and, as appropriate, be integrated with:

- The field operator's overall plan
- relevant external organisations, such as nearby field operators or national bodies.

Plans may require assessment and approval from local government authorities.

External bodies that may be involved in the emergency plan should be fully conversant with the appropriate parts of the plan and should participate in joint exercises and drills.

Gas tankers alongside the F(P)SO should be advised of the key elements of the F(P)SO's emergency plan as it relates to the ship, particularly the alarm signals and emergency shut down and departure procedures, details of which should be included in the Terminal Handbook.

Similarly, the F(P)SO needs to be aware of the gas tanker's actions in the event of an emergency onboard the gas tanker and the actions to be taken by any F(P)SO staff that may be onboard.

10.1.2.2 Control

The F(P)SO emergency plan should make absolutely clear the person or persons in charge who have overall responsibility for dealing with the emergency, listed in order of priority. Responsibilities for actions to be taken by others in the field to contain and control the emergency should also be clearly identified. Ultimate responsibility for an offshore emergency is likely to rest with the OIM of the F(P)SO.

Failure to define lines of responsibility can easily lead to confusion and to loss of valuable time.

10.1.2.3 Role of the Mooring Master

During cargo transfer operations, the Mooring Master will be on the gas tanker. In the event of an incident occurring on either the F(P)SO or the gas tanker, the Mooring Master's primary role should be one of liaison, providing communications between the gas tanker, tugs and the F(P)SO. The Mooring Master should have full knowledge of the F(P)SO's emergency response plan, should provide information to the F(P)SO's incident commander and provide advice to the decision on whether or not to remove the gas tanker.

The Mooring Master may have assistants and support crew on board the gas tanker who will provide back-up and assistance in the event of an emergency.

10.1.2.4 Role of the Gas Tanker's Master

The gas tanker's Master has the ultimate responsibility for the safety of his crew, his vessel and the environment. Neither the OIM of the F(P)SO nor the Mooring Master can override the Master's decisions to abort a berthing operation, stop cargo operations or vacate the berth if he feels the safety of his crew, his vessel or the environment may be compromised.

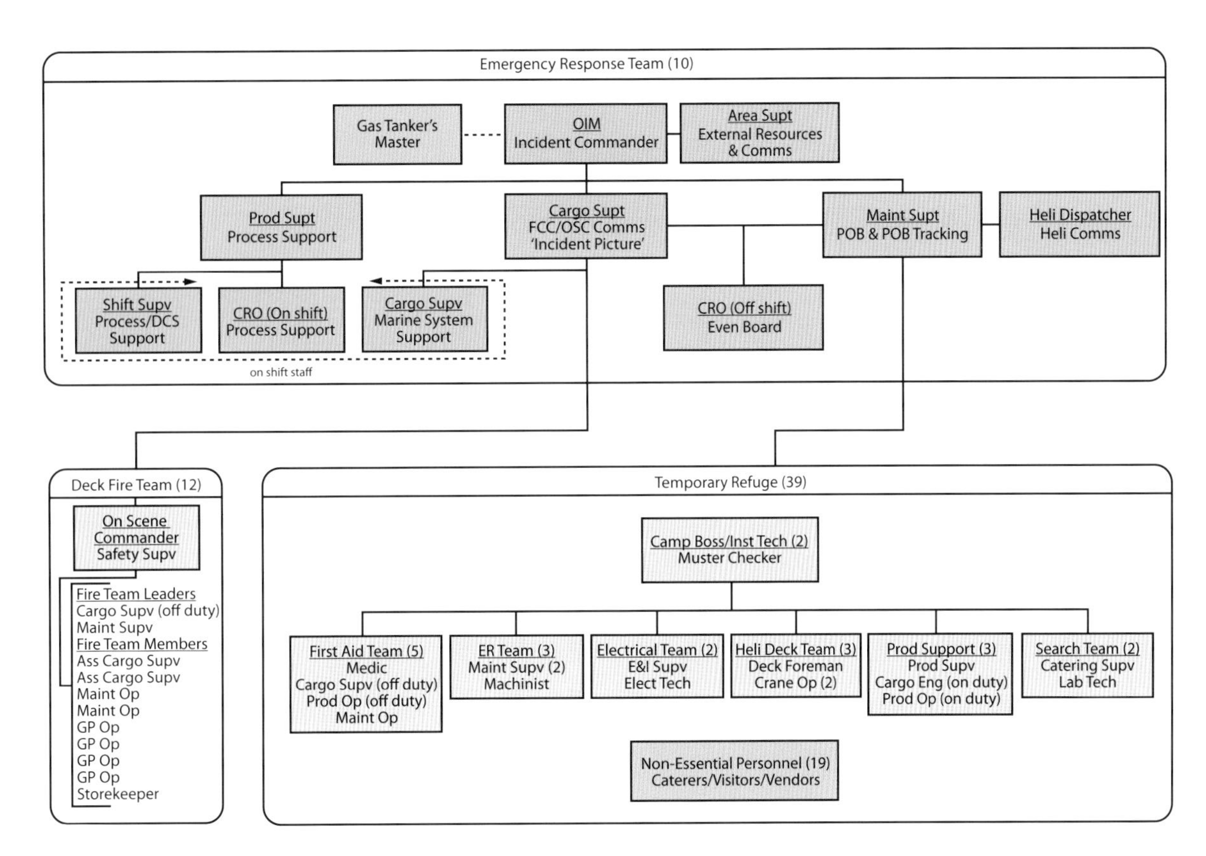

Figure 28: Example of an F(P)SO Emergency Response Organisation

10.1.3 Communications and Alarms

10.1.3.1 Alarms

The nature and type of alarms on the F(P)SO should be identified in the F(P)SO Terminal Handbook and the nature of the gas tanker's alarms should be advised to the F(P)SO. Conditioning and monitoring systems on gas installations may generate a number of audible alarms that are broadcast externally. An understanding of the meaning of the alarms and associated protocols should avoid confusion and undue concern.

There may be an option for a silent alarm, where only key personnel are alerted by radio, pager or hardwired telephone and put on alert. Typical applications would be in response to a security situation such as sabotage or bomb threat.

10.1.3.2 Contact Lists

The F(P)SO's emergency plan should include full contact details, both during and outside office hours, for those inside and outside the organisation who must be called in an emergency.

The names of alternates, who will be available in the event of the appointed person being absent or unavailable, should be included. Alternates should be fully aware of their responsibilities and be trained in the proper execution of their duties.

The contact list should be sufficiently comprehensive to eliminate the need to reference other documentation such as telephone directories.

The gas tanker should have a contact list for all personnel and organisations that may need to be contacted in an emergency.

10.1.3.3 Communication System Requirements.

Reliable communications are essential for dealing successfully with an emergency situation. Alternative power supplies should be provided if the primary source fails.

10.1.3.4 Communications Discipline

All personnel should understand and appreciate the necessity for strict observance of established rules of communications in an emergency, and should receive frequent instruction on the effective use of communications equipment and the procedures to follow.

Radio and telephone calls should be recorded and a log should be maintained in the control centre.

10.1.3.5 Field Plans and Charts

Charts or field plans should be available in the emergency control centre. While navigational charts may show the geographical location of the F(P)SO, platforms and other field infrastructure, they are not easily interpreted by non-mariners and should be supplemented with simple diagrammatic plans and schematics showing key installations, pipeline layouts and identifying nearby operators and interconnecting pipelines.

F(P)SO personnel should be familiar with the field plans and any issues relating to inter-dependency between fields, particularly the location and operation of production emergency shutdown systems.

10.1.3.6 Vessel Traffic Movement and Control

Emergency departure routes for the gas tanker should be pre-agreed and an individual should be assigned marine coordination duties in an emergency, to control all movements within the field.

10.1.3.7 Air Traffic and Control

Helicopter operations should be coordinated and controlled under normal operating conditions. In an emergency the need for good overall control is even more important.

10.1.3.8 Security Arrangements

It is recommended that F(P)SOs have security plans in line with 'ISPS Code' (Reference 15) regulations. Although international legislation may not require the F(P)SO to comply with ISPS requirements, lack of compliance may cause considerable problems for gas tankers at their next port of call.

Security arrangements should be appropriate for the prevailing conditions.

10.2 Emergency Scenarios

The emergency scenarios identified in Section 10.1 are briefly described below:

10.2.1 Power Failure on an F(P)SO, Gas Tanker or Tug

Despite all efforts to test propulsion systems and ensure adequate redundancy in power generation prior to and during berthing operations, power failures may occur for a number of reasons. F(P)SOs should give consideration to this possibility and identify actions to be carried out in the event of power failure:

- Prior to or during a berthing operation:

  The berthing operation should be aborted and the gas tanker removed to a safe anchorage or to an area clear of the F(P)SO. The berthing operation should not re-commence until the cause of the failure is determined to the satisfaction of the F(P)SO OIM and action taken to prevent re-occurrence

- during berthing with mooring lines being connected:

  Consideration should be given as to whether it is safer to continue mooring the gas tanker and secure it or to activate the emergency release of any moorings made fast and abort the operation, as above.

- loss of power on tug prior to or during berthing operation:

  Action to be taken will depend upon prevailing metocean conditions, which tug has failed, proximity to the F(P)SO at time of failure and whether mooring lines have been made fast.

10.2.2 Collision/Heavy Contact

Should the gas tanker land heavily against the fenders, with suspicion of damage, or make actual contact with the F(P)SO when berthing, cargo operations should not commence until an assessment of damage has been made. If necessary, the gas tanker should be moved off to validate the condition of the F(P)SO, gas tanker and equipment.

10.2.3 Grounding in Approaches to or within the Field

In the event of the gas tanker grounding on approach to the terminal or within the field of operation it should report the fact immediately to the F(P)SO or marine control and follow onboard contingency plans.

Navigational warnings should be issued and post-event the area of the grounding should be surveyed for obstructions, potential damage to subsea equipment and confirmation of water depths.

10.2.4 Loss of Containment

Any gas or liquid release should initiate a stop of the cargo transfer operation, which may be initiated automatically through gas detection or manually.

The source of the leak should be located and identified, the source isolated and action taken to effect a repair.

Hull protection or water deluge systems should be activated to reduce the risk of brittle fracture.

10.2.5 Uncontrolled Venting

A large gas release from either vessel should prompt the emergency suspension of operations and release of the gas tanker until normal conditions are re-instated.

During uncontrolled venting the decision to unmoor the export tanker should be made with due regard to the following:

- The presence of LPG vapours on deck and at the water surface about the vessels
- the hazards associated with the introduction of potential ignition sources, e.g. tugs
- sending personnel into a potentially explosive atmosphere should not be contemplated
- use of deluge systems on both vessels to knock down and disperse vapours
- use of thruster/heading control tug to orientate vessels into the best position to disperse vapours safely in prevailing wind.

A loss or reduction in capability of the reliquefaction plant on the F(P)SO may require the suspension of cargo operations. The cause of the failure should be determined and, if not recoverable and there is a risk of uncontrolled venting, the transfer should be stopped, hoses/arms disconnected and the gas tanker moved off to a safe distance clear of the F(P)SO.

10.2.6 Fire and Explosion

The discovery of a fire or risk of explosion on either vessel should prompt immediate suspension of operations and may lead to a decision to release the gas tanker from the F(P)SO.

10.2.7 Hose/Arm Failure

Any gas or liquid release from the transfer system should initiate a stop of the transfer operation, either automatically through gas detection or manually.

Hull protection or water deluge systems should be activated to reduce the risk of brittle fracture.

The source of the leak should be located and identified and, if required, the hose/arm should be cleared of product, disconnected and replaced.

10.2.8 Mooring Failure

Mooring tensions should be monitored, enabling trend patterns and peak loads to be observed and consideration to be given to stopping operations and disconnecting prior to any line failure. As with any inter-dependent system, the failure of a single element can result in progressive failures.

Loss of a single line may not be symptomatic of a system failure, but may be as a result of the individual line's poor condition. If progressive failures occur, a controlled shut down, hose/arm disconnection and unmooring should be initiated.

10.2.9 *Fender Breakaway or Burst*

Should a fender burst or break loose during berthing, the operation should be aborted and the gas tanker moved off to a safe location. The berthing operation should only recommence when fenders of the correct size and in serviceable condition have been reinstated.

A fender breaking loose or noted to be deflating during cargo operations could result in contact between the F(P)SO and gas tanker. The situation should be assessed and, if necessary, the cargo transfer operation halted and/or the gas tanker unmoored.

10.2.10 *Medical Evacuation*

Offshore installations such as F(P)SOs are generally equipped with medical facilities and qualified medical staff. These medical resources should be made available to the gas tanker should a crew member suffer an injury or serious illness.

The F(P)SO operator should be prepared to assist with arranging emergency medical evacuation (medivac) services in the event of a medical emergency onboard the gas tanker.

10.2.11 *Helideck Incident*

In the event of a helideck incident during a cargo transfer, all cargo operations should cease and emergency plans initiated. Depending on the circumstances, this may include emergency disconnection and removal of the gas tanker to a safe location.

10.2.12 *Oil Pollution*

In the event of oil being noted in the immediate vicinity of the gas tanker or the F(P)SO, this should be reported to the F(P)SO. Actions should be taken to immediately identify the source and stop the leak. Any spill response measures undertaken by the gas tanker or F(P)SO should only be initiated with the approval of the F(P)SO's OIM.

10.2.13 *Security Incident*

In the event of a breach in security or security threat to the F(P)SO or gas tanker while in the field, the F(P)SO or field operator should be immediately informed and should implement their security response plans.

Plans may require cargo operations to be suspended and the gas tanker removed to a safe location.

10.3 Drills and Training

It is important to distinguish between emergency drills and training. Training is used to impart knowledge and to gain practical experience. Drills are performed to prove the training and to exercise staff in a simulated emergency situation and to test the emergency response plans.

Drills and exercises should be based on emergency scenarios relevant to the operation and should be as realistic as possible. Where external groups are involved it will be necessary to produce a working document that all parties can review to ensure their own response is feasible and beneficial.

Drills and emergency exercises should be formally reviewed to enable their effectiveness to be assessed for the purpose of enhancing overall emergency preparedness.

The Safe Transfer of Liquefied Gas in an Offshore Environment

Appendices

A Example Content of Terminal Handbook (Port Information Book)

B Example of SIMOPS Matrix

C Example of Gas Tanker Nomination Questionnaire

D Checklists:

D1 Example Record of Pre-Transfer Meeting
D2 Example of F(P)SO Pre-Berthing Checklist
D3 Example of Ship/F(P)SO Safety Checklist
D4 Example Pre-Unberthing Checklist

Appendix A: Example Content of Terminal Handbook (Port Information Book)

The F(P)SO Regulations or Port Information Book should include details of the following:

- General description of the F(P)SO
- location of the F(P)SO, including applicable legislation and legislative requirements, observed time zone and relevant chart details
- F(P)SO-specific characteristics of the cargo
- personnel and equipment transfer procedures
- layout of the F(P)SO, including navigation aids, hazards to navigation, safety zones, and applicable anchoring/drifting areas
- communications
- emergency response
- F(P)SO facilities and services including tugs, bunkers, fuel, freshwater, stores, medical services, dental services, shore leave, repatriation, repairs, visitors and weather forecasting
- F(P)SO metocean conditions and environmental limitations on mooring/unmooring/cargo transfer operations
- fees and conditions of use of the F(P)SO facility
- environmental limitations including discharge of waste water, discharge of garbage/refuse, discharge of ballast water, boiler/soot release and reporting of environmental incidents
- safety limitations, including rules within F(P)SO zone, repairs and maintenance, hot work, ship-to-ship contacts and electric current, tank entry, alcohol and/or drugs, smoking, matches and lighters, photography/electronic equipment, swimming and fishing, arrival condition, anchors, main engine(s), emissions and potential F(P)SO ESD impacts, steering gear, boarding arrangements, accommodation requirements, use of radio and satellite communications equipment, use of radar and reporting of safety incidents
- minimum acceptance criteria of gas tankers, including standards, mooring equipment and arrangement, manifold equipment, lifting gear, personnel requirements and the vetting process
- notification of arrivals and communications
- mooring operations, including pilotage, inspection prior to mooring, testing of engines and steering gear, mooring process, cargo hose(s)/arm(s) connection process, ESD system requirements and testing, emergency towage and maintenance of mooring integrity
- cargo transfer operations, including safety check and pre-transfer meeting, commencement/continuation /completion process, helicopter operations, suspension of operations during electrical storms, hose/arm clearance, tank measurement/gauging and sampling
- unmooring operations, including disconnection of cargo hose(s)/arms, transfer of F(P)SO equipment from gas tanker to F(P)SO, unmooring process and early departure procedure

Documented procedures and information should include:

- Gas tanker nomination questionnaire
- gas tanker/F(P)SO safety operational agreement, including ship/F(P)SO safety check list, F(P)SO safety letter, HSE policy, security declaration, F(P)SO conditions, LPG material safety data sheets, fire and emergency instructions, and approved smoking areas
- gas tanker approach and mooring information, including Master/Mooring Master information exchange, mooring plan, use of emergency towing off pennants, fenders, tug information, and timesheet
- cargo transfer operation information, including loading plan, cargo manifold layout, safety checklist and note of protest pro-formas

- gas tanker mooring arrangement including fendering specifications
- cargo hose/arm connection process and hose/arm specifications
- tug and towage arrangement details, including policy on tugs remaining secured throughout operation
- F(P)SO ballast water management agreement.

Examples/Appendices may include the following:

- Diagram of field layout
- F(P)SO organisation chart
- ETA notices
- safety operational agreement
    - safety letter
    - declaration of security
    - F(P)SO Conditions
    - Material Safety Data Sheets
    - fire/emergency instructions
    - approved smoking area signs
    - incident reporting requirements
    - environmental reporting requirements
    - Master/Mooring Master information exchange
    - time sheet
    - cargo transfer plan
    - cargo manifold
    - gas tanker/F(P)SO safety checklist
    - notes of protest
- nomination questionnaire
- gas tanker approach and mooring
- gas tanker mooring arrangement
- gas tanker hose/arm connection
- tug and towage details
- ballast water management form
- personnel and equipment transfer arrangements
- SIMOPS.

Appendix B: Example SIMOPS Matrix

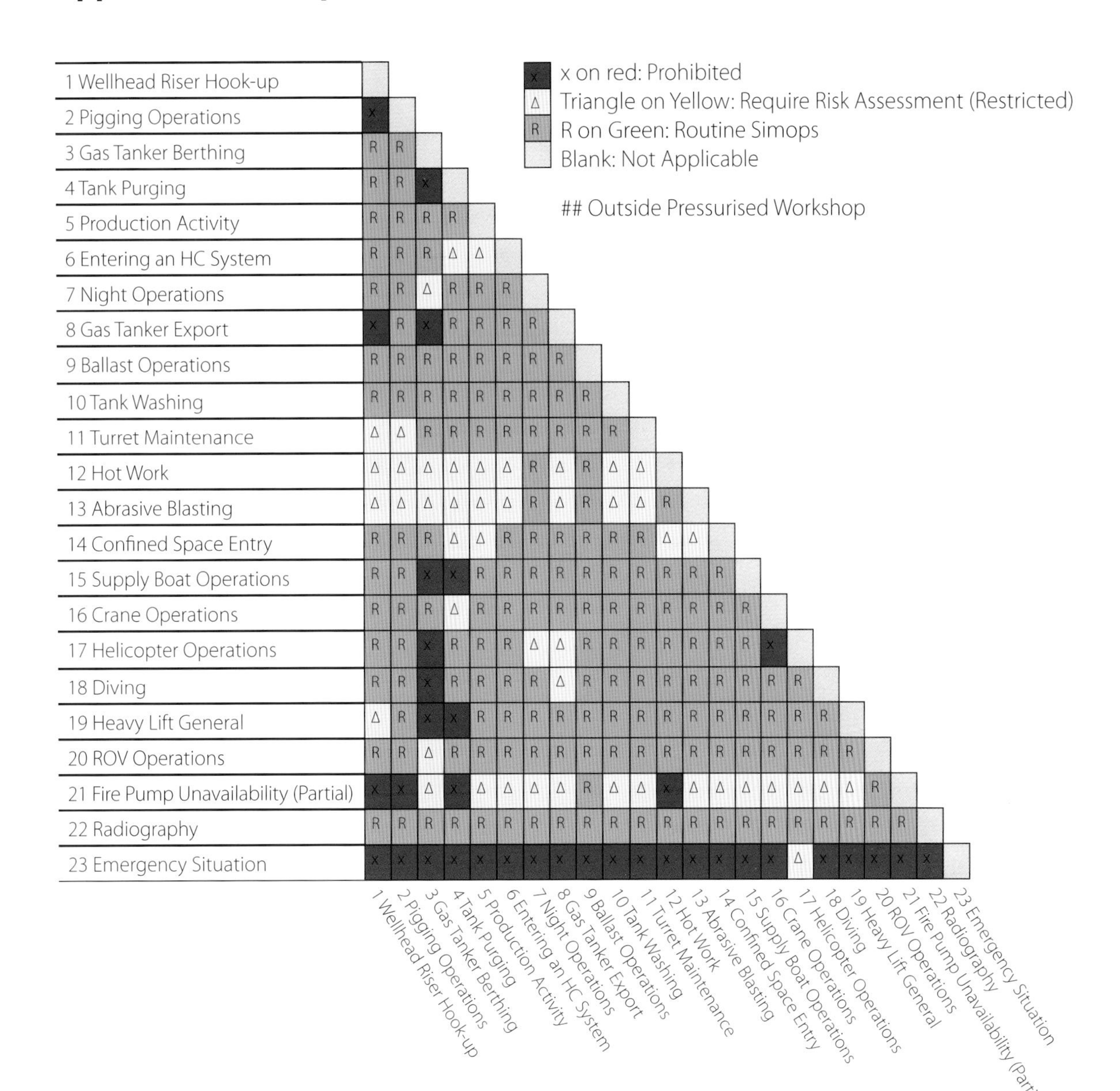

	1	2	3	4	5	6	7	8	9	10	11	12	13	14	15	16	17	18	19	20	21	22	23
1 Wellhead Riser Hook-up																							
2 Pigging Operations	x																						
3 Gas Tanker Berthing	R	R																					
4 Tank Purging	R	R	x																				
5 Production Activity	R	R	R	R																			
6 Entering an HC System	R	R	R	Δ	Δ																		
7 Night Operations	R	R	Δ	R	R	R																	
8 Gas Tanker Export	x	R	x	R	R	R	R																
9 Ballast Operations	R	R	R	R	R	R	R	R															
10 Tank Washing	R	R	R	R	R	R	R	R	R														
11 Turret Maintenance	Δ	Δ	R	R	R	R	R	R	R	R													
12 Hot Work	Δ	Δ	Δ	Δ	Δ	Δ	R	Δ	R	Δ	Δ												
13 Abrasive Blasting	Δ	Δ	Δ	Δ	Δ	Δ	R	Δ	R	Δ	Δ	R											
14 Confined Space Entry	R	R	R	Δ	Δ	R	R	R	R	R	R	Δ	Δ										
15 Supply Boat Operations	R	R	x	x	R	R	R	R	R	R	R	R	R	R									
16 Crane Operations	R	R	R	Δ	R	R	R	R	R	R	R	R	R	R	R								
17 Helicopter Operations	R	R	x	R	R	R	Δ	Δ	R	R	R	R	R	R	R	x							
18 Diving	R	R	x	R	R	R	R	Δ	R	R	R	R	R	R	R	R	R						
19 Heavy Lift General	Δ	R	x	x	R	R	R	R	R	R	R	R	R	R	R	R	R	R					
20 ROV Operations	R	R	Δ	R	R	R	R	R	R	R	R	R	R	R	R	R	R	R	R				
21 Fire Pump Unavailability (Partial)	x	x	Δ	x	Δ	Δ	Δ	Δ	R	Δ	Δ	x	Δ	Δ	Δ	Δ	Δ	Δ	Δ	R			
22 Radiography	R	R	R	R	R	R	R	R	R	R	R	R	R	R	R	R	R	R	R	R	R		
23 Emergency Situation	x	x	x	x	x	x	x	x	x	x	x	x	x	x	x	x	Δ	x	x	x	x	x	

x on red: Prohibited
Δ Triangle on Yellow: Require Risk Assessment (Restricted)
R on Green: Routine Simops
Blank: Not Applicable

Outside Pressurised Workshop

Appendix C: Example Gas Tanker Nomination Questionnaire

Guidance for Completion

The attached controlled questionnaire must be completed each time a gas tanker is nominated to the F(P)SO.

SWLs and dimensions are to be answered in metric units, as indicated.

Where a choice of answers is given, the answer(s) that do not apply should be deleted.

A legible copy of the gas tanker's mooring/general arrangement is to be attached to the completed questionnaire. The plan should depict the forecastle, main deck (including the manifold area) and poop deck locations of:

- Closed fairleads
- roller fairleads
- mooring bitts/bollards
- pedestal rollers
- winches and mooring lines.

Digital photographs of the above, including the portside manifold and the portside manifold tanker rail, are to be forwarded with this questionnaire.

The questionnaire, last three (3) cargo(s) certificates of quality prior to loading and the mooring/general arrangement plan are to be signed (name, position and company of signatory to be included) dated and marked with a company stamp to acknowledge that the information is accurate and all equipment is operational.

If there are subsequent changes in the validity of the information provided, or a change in the operational status of equipment, the F(P)SO should be advised at the earliest opportunity.

General Information

1	Name of vessel
2	Year of build
3	Flag and IMO number
4	Classification Society
5	Call sign
6	Type of vessel 1) Fully ref. 2)Semi ref. 3) Other
7	Last Three Cargoes
	1)
	2)
	3)
8	Cargo Capacity per tank at 98%

Tank No	M^3 at 15 deg C	MT w/density 0.506 at minus 42 deg C	MT w/density 0.574 at minus 5 deg C	Total in system	Total in MT
1					
2					
3					
4					

All above figures indicate total loadable quantities per cargo tank in Air.

Total Cargo:

9.	Is a copy of F(P)SO's Terminal Regulations on board	YES /NO
10.	If Yes, which edition	
11.	Is the owner/operator eligible to become, or is currently a member of SIGTTO.	YES /NO

Safety Management

12	Has vessel been involved in a pollution incident during the last 12 months?	YES /NO
13	Has vessel been involved in a grounding incident during the last 12 months?	YES /NO
14	Has vessel been involved in a collision during the last 12 months?	YES /NO
15	Has the vessel been inspected by an approved vetting/inspection authority within the last 12 months?	YES /NO

Vessel Management

16	Are all vessel documents and certificates valid, including the ISM Safety Management Certificate and the International Ship Security Certificate?	YES/ NO
17	Does the vessel comply with the ITF Agreement?	YES/ NO
18	Does the vessel have a current 1969 and 1992 Civil Liability Convention (CLC) certificate?	YES/ NO
19	Name of the P&I Club insuring the vessel	
20	Pollution liability insurance limit	USD
21	Does the vessel employ the OCIMF Drug and Alcohol policy or equivalent?	YES/ NO

Crew Management

22	Nationality of Officers	
23	Nationality of Crew	
24	Are manning levels in compliance with the minimum manning certificate?	YES/ NO
25	Are all officers and engineers qualified in accordance with STCW 95?	YES/ NO
26	Do the senior personnel speak English?	YES/ NO

Particulars

27	Length overall (LOA)	Metres
28	Extreme breadth	Metres
29	Distance bow – centre of manifold	Metres
30	Distance bridge front – centre of manifold	Metres
31	Freeboard in ballast condition	Metres
32	Freeboard in loaded condition	Metres
33	Height from ballast draught to bridge wing	Metres
34	Height from loaded draught to bridge wing	Metres
35	Ballast parallel mid body length – forward from centre of manifold	Metres
36	Ballast parallel mid body length – astern from centre of manifold	Metres
37	Loaded parallel mid body length – forward from centre of manifold	Metres
38	Loaded parallel mid body length – astern from centre of manifold	Metres
39	Transverse windage at ballast draught	m^2
40	Longitudinal windage at ballast draught	m^2
41	Summer Deadweight	Tonnes
42	Light Displacement	Tonnes
43	Loaded Displacement (Summer)	Tonnes

Manifold Equipment

44	Does the vessel comply with the latest edition of the applicable OCIMF /SIGTTO recommendations and have the capability to accept a load of 18 tonnes exerted by the F(P)SO's hose(s)?	YES/ NO
45	If NO, what does NOT comply?	
46	Are pressure gauges fitted outboard of manifold valves?	YES/ NO
47	Are temperature sensors fitted near the manifold and marked with minimum permitted cargo temperature?	YES/ NO
48	Can the vessel load (through separate lines) and carry two (2) fully refrigerated grades at the same time?	YES/ NO
49	Is butane loaded via forward port manifold and propane loaded by aft port manifold?	YES/ NO
50	Is a cross over(s) fitted at the manifold?	YES/ NO
51	Does the vessel have two 12 inch (304 mm) manifolds on the port side?	YES/ NO
52	If NO, Does the vessel have two 12 inch (304 mm) reducers in good condition?	YES/ NO
53	All reducers approved for minimum temperature of	°C
54	All reducers approved for maximum pressure of	ANSI psi
55	Horizontal distance from vessel's vertical side to the presentation flanges of the port side manifold.	Metres
56	Are there at least two (2) cruciform bollards located either between the manifold or one forward and one aft of the manifold area?	YES/ NO
57	Is the SWL of these manifold cruciform bollards twenty five (25) tonnes minimum?	YES/ NO
58	Is one (1) closed chock fairlead fitted for use with each manifold cruciform bollard with a clear opening of four hundred (400) x two hundred and fifty (250) millimetres?	YES/ NO
59	Are there eye plates located in the immediate vicinity of each manifold?	YES/ NO
60	Is the SWL of these manifold located eye plates fifteen (15) tonne minimum?	YES/ NO
61	Are 'full size' mooring bitts and panama chocks fitted within thirty five (35) metres of the centre of the manifold, fore and aft?	YES/ NO
62	Manifold construction material	Carbon steel/non magnetic Stainless steel
63	Type of ESD valve	Butterfly/ball/ gate
64	Diameter of ESD valve	mm
65	Time from ESD initiation to valve(s) fully closed	secs
66	Time of valve closure from start of valve movement to fully closed	secs
67	Can the vessel adjust the time in (65) above?	YES/ NO
68	Can the vessel adjust the time in (66) above?	YES/ NO
69	Is the vessel fitted with SIGTTO ESD link system?	YES/ NO

70	Advise all possible tank configuration option basis: Simultaneous Loading, i.e. to load two fully refrigerated grades at the same time through separate lines. Butane loaded via forward port manifold and propane loaded via aft port manifold. Option 1) Option 2) Option 3) Option 4) Option 5)	YES/ NO
71	Total number of crew onboard the vessel	_ persons
72	Can the vessel accommodate 4 to 5 Terminal Mooring Personnel?	YES/ NO
73	Is the lifeboat capacity sufficient to accommodate all crew, including terminal mooring personnel?	YES/ NO

Cargo Manifold Layout

Please complete:

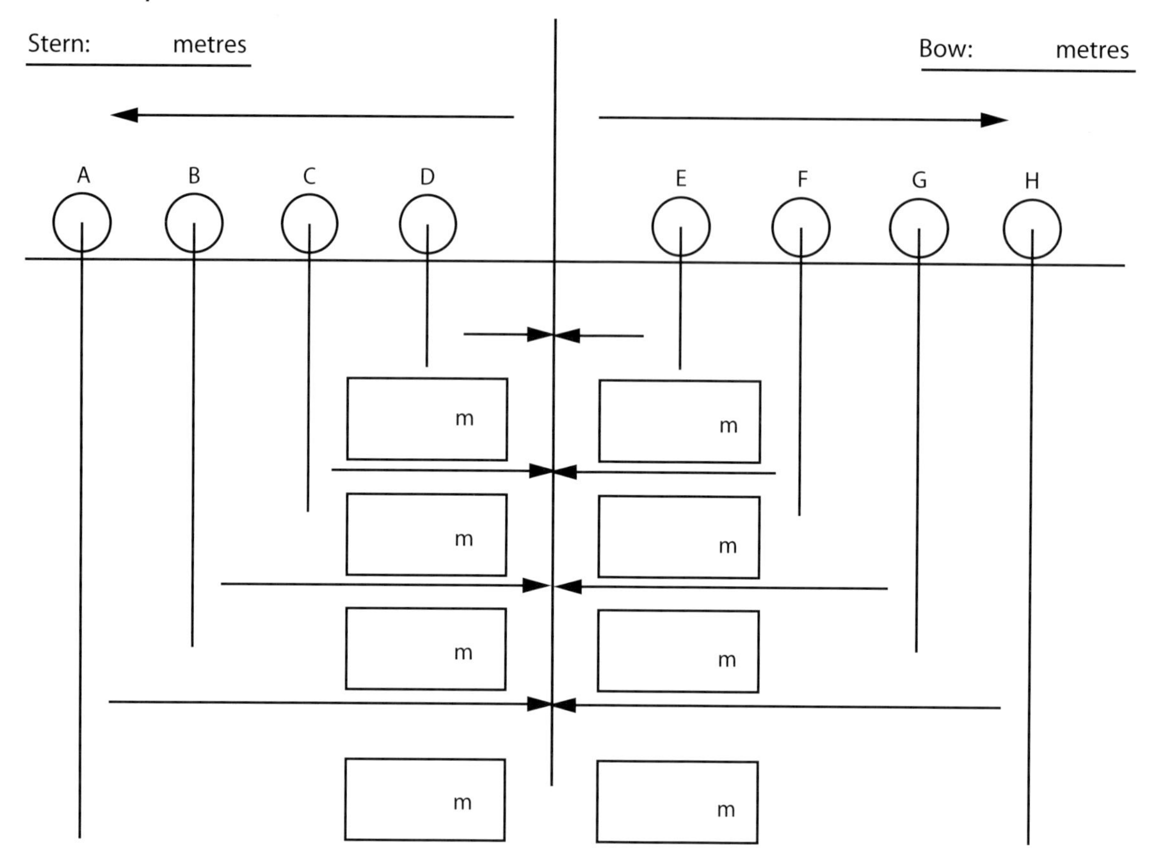

Please complete:

Pipe Flange	A	B	C	D	E	F	G	H
Duty (Bunker/ Vapour/ Liquid/ N_2)								
Rating/ Class (ANSI psi)								
Size (mm)								
Raised or Flat face								
Flange Thickness (mm)								
Height above Main Deck (m)								
Height above Drip Tray (m)								
Height above Loaded Draught (m)								
Height above Ballast Draught (m)								
Horizontal Distance Rail to Manifold (m)								

Reducers

Please complete:

Size (mm) (ANSI Class)	To	Size (mm) (ANSI Class)	Qty	Liquid/ Vapour/ Both	Remarks
300				L/ V / B	
300				L/ V / B	
300				L/ V / B	
300				L/ V / B	
300				L/ V / B	

150				L/ V / B	
150				L/ V / B	
150				L/ V / B	
150				L/ V / B	
150				L/ V / B	

Cargo Tank Equipment

74	Will the gas tanker arrive with cargo tanks cold and 'gassed up' ready to load propane and butane immediately?	YES/ NO
75	If No, additional cooldown time required	hours
76	Is vessel fitted with deck storage tanks?	YES/ NO
77	If Yes, state contents	
78	State any 'sloshing' restrictions	
79	Minimum working pressure	
80	Maximum working pressure	
81	Maximum allowable cargo tank relief valve settings	bar
82	Minimum acceptable temperature	°C

Loading Rates

Please complete:

	Propane (m^3/hour)	Butane (m^3/hour)	Remarks
Density			
Without vapour return			
@ minus 42 °C		N/A	
@ minus 5 °C	N/A		

83	All reliquefaction equipment in demonstrable working order able to handle the total vapour evolved during the cargo loading operation and be capable of maintaining the product at the same temperature at which it was received at the gas tanker's loading manifold or at the gas tanker's minimum tank temperature, whichever is the higher.	YES /NO

Lifting Equipment

84	SWL of port side crane/derrick	MT
85	If derrick, is it able to slew and top under power while under load	YES /NO / NA
86	Is safe working load (SWL) clearly marked on all lifting equipment and are all test certificates available on board	YES/ NO
87	What is the maximum outreach of vessel's cranes or derricks outboard of the vessel's side	m

Mooring Line Equipment

Please complete:

Mooring Lines on Drums	No	Type	Diameter (mm)	Length (m)	MBL (T)	Closed Fairlead Size (mm x mm)	Soft Rope Tails Length & MBL (T)	Remarks
Forecastle								
Fwd Main Deck								
Aft Main Deck								
Poop								
Other								
Emerg'cy Tow Wires						N/A		

Please complete:

Mooring Lines on Drums	No	Type (Single or Double Drums)	Split Drum	Steam or Hyd	Heaving Power (T)	Brake SWL & Render Load (T)	Date of last Test	Hauling Speed (m/min)
Forecastle								
Fwd Main Deck								
Aft Main Deck								
Poop								
Other								
Emerg'cy Tow Wires								

88	Does the vessel comply with the latest edition of the ICS/OCIMF/SIGTTO 'Ship to Ship Transfer Guide (Liquefied Gases)'?	YES/ NO
89	Does the vessel comply with the latest edition of the OCIMF 'Mooring Equipment Guidelines'?	YES/ NO
90	Do all mooring lines pass through closed – not open – fairleads?	YES/ NO

Signature .... Name ....

Position .... Company ....

Date ....

Appendix D1: Example Record of Pre-Transfer Meeting

<table>
<tr><th>No</th><th colspan="3">Sequence of Operations</th><th>Remarks</th></tr>
<tr><td>1</td><td colspan="3">Safety</td><td></td></tr>
<tr><td></td><td colspan="3">• Safety Checklist completed by Mooring Master and ship staff
• weather forecast
• adverse weather checklist.</td><td>'R' Items re-check every 12 hrs by gas tanker and F(P)SO staff.</td></tr>
<tr><td>2</td><td colspan="3">Communications</td><td></td></tr>
<tr><td></td><td colspan="3">• See F(P)SO regulations
• primary comms VHF 7A, 16, 90A + 91
• emergency signal by VHF and F(P)SO/Gas tanker's whistle (7 short + 1 long blast)</td><td>VHF Ch 7A is preferred channel for cargo operations</td></tr>
<tr><td>3</td><td colspan="3">Operations</td><td></td></tr>
<tr><td>3.1</td><td colspan="3">Liquid hose connection – then vapour return hose gas tanker's crew to assist with use of deck crane, removing blanks, etc.

Mooring Master & 2 assistants required for basket transfer.</td><td>Starboard side (FPSO)
1 x 12" (150) Liquid, 1 x 12"(150) Liquid,
1 x 8" (150) Vapour.
Port side (FPSO) - 1 x 6" (300) Liquid
Hoses to be monitored hourly.
• FPSO crew to connect hose to manifold with gas tanker's crew assistance</td></tr>
<tr><td>3.2</td><td colspan="3">Hose pressure/leak test to Max 6 Barg using hot gas from reliq compressors.</td><td></td></tr>
<tr><td>3.3</td><td colspan="3">Hose cooldown & gas tanker Max Rate
Max Permitted Full Rate C_3..............................
Max Permitted Full Rate C_4..............................</td><td>Initial slow rate as per gas tanker's requirements?
....
Hose to be fully cooled down before maximum rate (Approx. minus 30° (C_3)) (Approx 0° (Ref C_4))</td></tr>
<tr><td rowspan="7">3.4</td><td>LPG cargo loading: Indicate quantity in each tank</td><td>Tank</td><td>Product/ MT</td><td rowspan="7">• Rate increased by 1 x 550 m^3/hr pump at a time.
• C_3 Total 6 Pumps Available.
• C_4 Total 2 Pumps Available.
• Target rate C_4 1,100 m^3/hr, C_3 2,800 m^3/h
• Sampling to commence when at full rate.</td></tr>
<tr><td></td><td>Tk.1</td><td></td></tr>
<tr><td></td><td>Tk.2</td><td></td></tr>
<tr><td></td><td>Tk.3</td><td></td></tr>
<tr><td></td><td>Tk.4</td><td></td></tr>
<tr><td></td><td>Tk.5</td><td>N/A</td></tr>
<tr><td></td><td>Tk.6</td><td>N/A</td></tr>
<tr><td>3.5</td><td colspan="3">Cargo tank topping off</td><td>• Surveyor to be advised 1 hr. before stop
• loading pumps stopped one at a time
• final completion with 1 pump only for minimum 10 min.</td></tr>
<tr><td>3.6</td><td colspan="3">Loading hoses draining</td><td>• Hoses will be blown with hot LPG vapour to gas tanker from F(P)SO
• when hose clear of liquid, manifold end valve to be closed</td></tr>
<tr><td>3.7</td><td colspan="3">Loading hose disconnection</td><td>• F(P)SO crew to disconnect. Assistance required from gas tanker's crew.</td></tr>
<tr><td>3.8</td><td colspan="4">Comments:
Schedule
Safety Check-List</td></tr>
</table>

Appendix D2: Example of F(P)SO Pre-Berthing Checklist

F(P)SO is to confirm its readiness, via radio to the Mooring Master prior to berthing the gas tanker, as per this checklist.

No.	Item to be checked	Yes/No	Remarks
1	Are primary/secondary fenders in place with fender pennants and in good condition?		
2	Have over-side protrusions on the side of berthing been retracted?		
3	Are all vessels engaged in providing stores/bunkers to the F(P)SO cast off and is the heliport closed during berthing operations?		
4	Is the thruster operational?		
5	Are the crew briefed on the mooring procedure and the mooring line sequence?		
6	Are mooring ropes, rope messengers, rope stoppers and heaving lines ready for use?		
7	Is power on for the winches and are they in good order.		
8	Are the pneumatic line throwers charged and ready for use?		
9	Are the quick release hooks reset and ready for use?		
10	Are the crew standing by at the mooring stations and communications established?		
11	Has the hose handling crane been tested and ready for operation?		
12	Is the F(P)SO in all respects ready for the intended cargo transfer operation?		

Date and Time Checklist completed ....

Appendix D3: Example of Ship/ F(P)SO Safety Checklist

Part 'A' – Bulk Liquid General – Physical Checks

Bulk Liquid – General	Ship	Terminal	Code	Remarks
1. There is safe access between the ship and F(P)SO.			R	
2. The ship is securely moored.			R	
3. The agreed ship/shore communication system is operative.			A R	
4. Emergency towing-off pennants are correctly rigged and positioned.			R	
5. The ship's fire hoses and fire fighting equipment are positioned and ready for immediate use.			R	
6. The F(P)SO's fire fighting equipment is positioned and ready for immediate use.				
7. The ship's cargo and bunker hoses, pipelines and manifolds are in good condition, properly rigged and appropriate for the service intended.				
8. The F(P)SO's cargo and bunker hoses/arms are in good condition, properly rigged and appropriate for the service intended.				
9. The cargo transfer system is sufficiently isolated and drained to allow safe removal of blank flanges prior to connection.				
10. Scuppers and save-alls onboard are effectively plugged and drip trays are in position and empty.			R	
11. Temporarily removed scupper plugs will be constantly monitored.			R	
12. F(P)SO spill containment and sumps are correctly managed.			R	
13. The ship's unused cargo and bunker connections are properly secured with blank flanges fully bolted.				

14. The F(P)SO's unused cargo and bunker connections are properly secured with blank flanges fully bolted.				
15. All cargo, ballast and bunker tank lids are closed.				
16. Sea and overboard discharge valves, when not in use, are closed and visibly secured.				
17. All external doors, ports and windows in the accommodation, stores and machinery spaces are closed. Engine room vents may be open.			R	
18. The ship's emergency fire control plans are located externally.				Location ....

Part 'B' – Bulk Liquid General – Verbal Verification

Bulk Liquid – General	Ship	Terminal	Code	Remarks
19. The ship is ready to move under its own power.			P R	
20. There is an effective deck watch in attendance onboard and adequate supervision of operations on the ship and the F(P)SO.			R	
21. There are sufficient personnel onboard and on the F(P)SO to deal with an emergency.			R	
22. The procedures for cargo and ballast handling have been agreed			A R	
23. The emergency signal and shutdown procedure to be used by the ship and F(P)SO have been explained and understood			A	
24. Material Safety Data Sheets (MSDS) for the cargo transfer have been exchanged where requested.			P R	
25. The hazards associated with toxic substances in the cargo being handled have been identified and understood.				
26. The agreed tank venting system will be used.			A R	Method ....
27. The requirements for closed operations have been agreed.			R	
28. The operation of the P/V system has been verified.				
29. Where a vapour return line is connected, operating parameters have been agreed.			A R	
30. Independent high level alarms, if fitted, are operational and have been tested.				
31. Adequate electrical insulating means are in place in the transfer connection.				
32. The F(P)SO lines are fitted with a non-return valve or procedures to avoid back filling have been discussed.				
33. Smoking rooms have been identified and smoking requirements are being observed.			A R	Nominated smoking rooms:

34. Naked light regulations are being observed.			A R	
35. Telephones, mobile phones and pager requirements are being observed.			A R	
36. Hand torches (flashlights) are of an approved type.				
37. Fixed VHF/UHF transceivers and AIS equipment are on the correct power mode or switched off.				
38. Portable VHF/UHF transceivers are of an approved type.				
39. The ship's main radio transmitter aerials are earthed and radars are switched off.				
40. Electric cables to portable electrical equipment within the hazardous area are disconnected from power.				
41. Window type air conditioning units are disconnected.				
42. Positive pressure is being maintained inside the accommodation, and air conditioning intakes, which may permit the entry of cargo vapours, are closed.				
43. Measures have been taken to ensure sufficient mechanical ventilation in the pumproom.			R	
44. There is provision for an emergency escape.				
45. The maximum wind and swell criteria for operations has been agreed.			A	Stop cargo at: .... Disconnect at: .... Unberth at: ....
46. Security protocols have been agreed between the Ship Security Officer and the F(P)SO Security Officer, if appropriate.			A	
47. Where appropriate, procedures have been agreed for receiving nitrogen supplied from the F(P)SO, either for inerting or purging ship's tanks, or for line clearing into the ship.			A P	

Part 'D' – Bulk Liquefied Gases – Verbal Verification

Bulk Liquid – General	Ship	Terminal	Code	Remarks
1. Material Safety Data Sheets are available giving the necessary data for the safe handling of the cargo.				
2. A manufacturer's inhibition certificate, where applicable, has been provided.			P	
3. The water spray system is ready for immediate use.				
4. There is sufficient protective equipment (including self-contained breathing apparatus) and protective clothing ready for immediate use.				
5. Hold and inter-barrier spaces are properly inerted or filled with dry air, as required.				
6. All remote control valves are in working order.				
7. The required cargo pumps and compressors are in good order, and the maximum working pressures have been agreed between ship and shore.			A	
8. Re-liquefaction or boil off control equipment is in good order.				
9. The gas detection equipment has been properly set for the cargo, is calibrated, has been tested and inspected and is in good order.				
10. Cargo system gauges and alarms are correctly set and in good order.				
11. Emergency shutdown systems have been tested and are working properly.				
12. Ship and F(P)SO have informed each other of the closing rate of ESD valves, automatic valves or similar devices.			A	Ship .... Shore ....
13. Information has been exchanged between ship and F(P)SO on the maximum/ minimum temperatures/ pressures of the cargo to be handled.			A	

14. Cargo tanks are protected against inadvertent overfilling at all times while any cargo operations are in progress.				
15. The compressor room is properly ventilated; the electrical motor room is properly pressurised and the alarm system is working.				
16. Cargo tank relief valves are set correctly and actual relief valve settings are clearly and visibly displayed. (Record settings below.)				

Tank No 1		Tank No 5		Tank No 8	
Tank No 2		Tank No 6		Tank No 9	
Tank No 3		Tank No 7		Tank No 10	
Tank No 4					

We, the undersigned, have checked the above items in Parts A, B and Part D, in accordance with the instructions and have satisfied ourselves that the entries we have made are correct to the best of our knowledge.

We have also made arrangements to carry out repetitive checks as necessary and agreed that those items with code 'R' in the Checklist should be re-checked at intervals not exceeding 12 hours.

If to our knowledge the status of any item changes, we will immediately inform the other party.

For Ship	For F(P)SO
Name..	Name..
Rank..	Rank..
Signature..	Signature..
Date..	Date..
Time..	Time..

The presence of the letters 'A', 'P' or 'R' in the column entitled 'Code' indicates the following:

A ('Agreement'). This indicates an agreement or procedure that should be identified in the 'Remarks' column of the Check-List or communicated in some other mutually acceptable form.

P ('Permission'). In the case of a negative answer to the statements coded 'P', operations should not be conducted without the written permission from the appropriate authority.

R ('Re-check'). This indicates items to be re-checked at appropriate intervals, as agreed between both parties, at periods stated in the declaration.

Appendix D4: Example Pre-Unberthing Checklist

Items to be checked	Yes/No	Remarks
1. Are cargo hoses properly purged prior to hose disconnection?		
2. Are cargo hose ends and manifolds blanked?		
3. Is the transfer side of both vessels clear of obstructions and are cranes/derricks stowed?		
4. Are fenders, including fender pennants, in good order?		
5. Are the navigational, propulsion and steering systems ready in all respects on the gas tanker?		
6. Has the method of unberthing and sequence of letting go of moorings been agreed between the gas tanker and the F(P)SO?		
7. Has the crew been briefed on the un-mooring procedure?		
8. Is the power on for the winches and windlass and are they in good order?		
9. Are crews standing by at their mooring stations and have communications been tested between F(P)SO and the gas tanker?		
10.Are sufficient crew available for simultaneously letting go the gas tanker fore & aft and are the respective bridges manned?		
11.If required, have the crews been instructed on the use of the quick release hooks and to let go only as requested by the Mooring Master on the gas tanker?		
12.Is adequate lighting available, particularly over the side in the vicinity of the fenders?		
13.Has the prevailing weather and current data and actual conditions been assessed?		
14.Has the correct heading of F(P)SO been determined for unberthing operations?		
15.Are the tugs secured to the gas tanker and communications tested?		
16.Has the shipping traffic in the vicinity been checked and advised of the intended manoeuvre?		
17.Has the Heliport on the F(P)SO been closed for the un-berthing operations?		

Date and Time Checklist completed ....